Photo-Electrochemical Ammonia Synthesis

Emerging Materials and Technologies

Series Editor

Boris I. Kharissov

Recycled Ceramics in Sustainable Concrete: Properties and Performance

Kwok Wei Shah and Ghasan Fahim Huseien

Photo-Electrochemical Ammonia Synthesis: Nanocatalyst Discovery, Reactor Design, and Advanced Spectroscopy

Mohammadreza Nazemi and Mostafa A. El-Sayed

Fire-Resistant Paper: Materials, Technologies, and Applications

Ying-Jie Zhu

Sensors for Stretchable Electronics in Nanotechnology

Kaushik Pal

Polymer-based Composites: Design, Manufacturing, and Applications

V. Arumugaprabu, R. Deepak Joel Johnson, M. Uthayakumar, and P. Sivaranjana

Nanomaterials in Bionanotechnology: Fundamentals and Applications

Ravindra Pratap Singh and Kshitij RB Singh

Photo-Electrochemical Ammonia Synthesis

Nanocatalyst Discovery, Reactor Design, and Advanced Spectroscopy

Mohammadreza Nazemi
Mostafa A. El-Sayed

CRC Press
Taylor & Francis Group
Boca Raton London New York

CRC Press is an imprint of the
Taylor & Francis Group, an **informa** business

First edition published 2022
by CRC Press
6000 Broken Sound Parkway NW, Suite 300, Boca Raton, FL 33487-2742

and by CRC Press
2 Park Square, Milton Park, Abingdon, Oxon, OX14 4RN

CRC Press is an imprint of Taylor & Francis Group, LLC

Library of Congress Cataloging-in-Publication Data
Names: Nazemi, Mohammadreza, author. | El-Sayed, Mustafa A., author.
Title: Photo-electrochemical ammonia synthesis : nanocatalyst discovery, reactor design, and advanced spectroscopy / authored by Mohammadreza Nazemi and Mostafa A. El-Sayed.
Description: First edition. | Boca Raton, FL : CRC Press, [2021] | Series: Emerging materials and technologies | Includes bibliographical references. | Summary: "This book covers the synthesis of novel hybrid plasmonic nanomaterials and their application in photo-electrochemical systems to convert low energy molecules to high value-added molecules and looks specifically at photo-electrochemical nitrogen reduction reaction (NRR) for ammonia synthesis as an attractive alternative to the long-lasting thermochemical process. This book will be of high interest to researchers, advanced students, and industry professionals in working in sustainable energy storage and conversion across the disciplines of Chemical Engineering, Mechanical Engineering, Materials Science and Engineering, Environmental Engineering, and related areas"-- Provided by publisher.

Identifiers: LCCN 2021009044 (print) | LCCN 2021009045 (ebook) | ISBN 9780367694371 (hbk.) | ISBN 9780367694388 (pbk) | ISBN 9781003141808 (ebk)
Subjects: LCSH: Ammonia--Synthesis. | Photoelectrochemistry.
Classification: LCC TP223 .N39 2021 (print) | LCC TP223 (ebook) | DDC 661/.34--dc23
LC record available at https://lccn.loc.gov/2021009044
LC ebook record available at https://lccn.loc.gov/2021009045

ISBN: 978-0-367-69437-1 (hbk)
ISBN: 978-0-367-69438-8 (pbk)
ISBN: 978-1-003-14180-8 (ebk)

Typeset in Times
by SPi Global, India

Contents

Notes on the Authors

Mohammadreza Nazemi is a postdoctoral fellow in the School of Chemistry and Biochemistry at the Georgia Institute of Technology (Georgia Tech). He received his Ph.D. from the Woodruff School of Mechanical Engineering at Georgia Tech under the supervision of Prof. Mostafa El-Sayed in 2020. He received his BS degree (2013) in Aerospace Engineering from the Sharif University of Technology and MS degree (2015) in Mechanical Engineering from Michigan Technological University. His current research focuses on the development and testing of hollow plasmonic nanostructures for photo-electrochemical energy generation. In addition, he is using advanced spectroscopic and microscopic techniques to gain a mechanistic understanding of photo-electrochemical (de) hydrogenation reactions for sustainable fuel and fertilizer production.

Mostafa A. El-Sayed is the Regents' Emeritus Professor in the School of Chemistry and Biochemistry at the Georgia Institute of Technology (Georgia Tech). He obtained his Ph.D. from Florida State University in 1959 with Michael Kasha, and after postdoctoral fellowships at Harvard, Yale, and Caltech, he joined the faculty of School of Chemistry and Biochemistry at University of California, Los Angeles (UCLA) in 1961 and Georgia Tech in 1994. He is currently an elected member of the U.S. National Academy of Science, an elected fellow of the American Academy of Arts and Sciences, a fellow of the American Association for the Advancement of Science (AAAS) and of the American Physical Society (APS), and former editor-in-chief of the Journal of Physical Chemistry. He is the recipient of several prestigious awards including ACS Priestly medal, Ahmed Zewail prize in molecular sciences, the ACS Irving Langmuir Prize in Chemical Physics, the Glenn T. Seaborg Medal, and the U.S. National Medal of Science. He was included in the top 1% most cited researchers in 2017 and 2018 (web of science).

Preface

Satisfying global energy needs in an environmentally friendly manner by harvesting renewable energy sources is a grand challenge that we face today and has become increasingly urgent. As renewable energy sources are either intermittent in nature or remote in location, there is a need to develop cost-effective and sustainable methods of storing and transporting this energy on an industrial scale. Currently, the chemical industry heavily relies on fossil feedstock as energy sources. As the cost of renewably derived electricity continues to decrease, electrically driven approaches are a promising technique for the synthesis of fuels and high-value chemicals in a clean, sustainable, and distributed manner. Electrosynthesis enables the use of renewable electricity to convert low energy molecules (e.g., H_2O, N_2) to value-added molecules (e.g., H_2, NH_3) that can be utilized as either fuel, energy storers, or chemicals.

This book will cover the synthesis of novel hybrid plasmonic nanomaterials and their applications in photo-electrochemical systems to convert low energy molecules (N_2) to high-value-added molecules (NH_3). *Operando* spectroscopy and density functional theory (DFT) calculations will be discussed to identify intermediate species relevant to the catalytic reactions at the electrode–electrolyte interfaces to provide insight into the reaction mechanism.

Photoelectrochemical nitrogen reduction reaction (NRR) for ammonia synthesis provides an attractive alternative to the long-lasting thermochemical Haber–Bosch process, in a clean, sustainable, and decentralized way, if the process is coupled to renewably derived electricity sources. Ammonia has a widespread utility as a precursor for making fertilizer. Ammonia also holds great promise as a carbon-neutral liquid fuel for storing intermittent renewable energy sources as well as for power generation due to the compound's high energy density (5.6 MWh ton^{-1}) and hydrogen content (17.6 wt.%). Electrification of ammonia synthesis on a large scale requires an effective electrocatalyst that converts N_2 to NH_3 with a high yield and efficiency. The selectivity of N_2 molecules on the surface of the catalyst has been demonstrated to be one of the significant challenges in enhancing the rate of photo-electrochemical NRR in an aqueous solution under ambient conditions. The rational design of electrode–electrolyte and the reactor in the context of photo-electrochemical systems is required to overcome the selectivity and activity barrier in the nitrogen reduction reaction. A range of topics will be discussed in this book to address a critical obstacle to achieving the overarching goal of distributed ammonia synthesis.

This book aims to provide an integrated scientific framework to overcome the challenges of renewable energy storage and transport. It also contributes to future electrification, decarbonization, and sustainability of the modern chemical industry.

Acknowledgements

The authors acknowledge that the material published in this book is based upon work supported by the National Science Foundation under grant no. 1904351. This work was performed in part at the Georgia Tech Institute for Electronics and Nanotechnology, a member of the National Nanotechnology Coordinated Infrastructure (NNCI), which is supported by the National Science Foundation (grant ECCS-1542174).

We wish to take this opportunity to thank all those who have contributed to this work.

Profs. Younan Xia, Meilin Liu, Paul Kohl, Thomas Orlando, Vladimir Tsukruk, Todd Sulchek, Matthew McDowell, and Ting Zhu for providing constructive feedback and suggestions. In addition, we are grateful to Dr. Sajanlal Panikkanvalappil, Dr. Pengfei Ou, Dr. Jing Zhou, Dr. Shuaidi Zhang, Dr. Saewon Kang, Dr. Ali Abdelhafiz, Luke Soule, Abdulaziz Alabbady, Jalen Borne and Alan Liu for their help and input on various aspects.

We gratefully acknowledge all staff members at the Institute for Electronics and Nanotechnology at Georgia Tech. In particular, we thank Walter Henderson, Eric Woods, Todd Walters, David Tavakoli, Rathi Monikandan, and Yong Ding. Their guidance and support to operate all instruments were incredible, which enabled us to achieve meaningful results on time. Thanks also to Dr. Leslie Gelbaum for helping us with NMR measurements, and Ms. Tabassum Shah for performing ICPES measurements.

Additionally, we acknowledge our collaborators in the VentureLab and Scheller College of Business at Georgia Tech, including Mr. Jonathan Goldman, Prof. Jonathan Giuliano, Prof. Nicole Morris, Prof. Robert Gemmell, and Mr. Trevor Brown from Ammonia Energy Association for their help and feedback on the commercialization of renewable ammonia synthesis. We also appreciate the contributions of MBA students Pete Nicolay, Ian Schubert, Michael Luo from the Scheller College of Business at Georgia Tech, and JD students Amanda Guarisco, Aaron Savit from the Emory School of Law on this work.

We are very grateful to our lab administrative support, Ms. Michele Yager, not only for ensuring our administrative work always ran smoothly and without any delay, but also for always being available to give us her best advice. We would also like to thank Mr. Mike Riley for his assistance in making sure our lab was safe and functional.

Last but not least, we would like to thank our families for their endless love and support. We dedicate this book to you.

1 Ammonia
A Multi-Purpose Chemical

1.1 INTRODUCTION: NANOCATALYSIS USING METALLIC AND SEMICONDUCTOR NANOPARTICLES

Nanocatalysis has attracted great attention in the past two decades in homogeneous solution-phase colloidal reactions and heterogeneous supported nanoparticle gas-phase reactions. Nanoparticles are characterized by a high surface-to-volume ratio, resulting in increased efficiency in catalysis.[1–6] Introducing sharp corners, edges, and defect sites to the nanocatalysts further increase their catalytic efficiency.[7] Since the activity of the nanocatalyst depends on the number of high energy active centers,[8] atoms located at the sharp tips are thermodynamically and catalytically more active because they are unsaturated in their chemical valency (i.e., atoms that do not have the complete number of bonds that they can chemically accommodate to act as active sites than smoother nanoparticles).[6,9–11] Sharp tips are prone to be rounded during the catalytic reactions. This causes a remarkable decrease in their efficacy. The shape change of the particles could be due to leaching or rearrangement of the surface atoms.[9,12]

There are two main categories of nanoparticles: solid and hollow. In nanocatalysis by solid nanoparticles, the catalytic reaction happens by involving the atoms from the surface of the nanocatalyst, while for hollow nanocatalysts the reaction takes place at the outer and/or the inner surfaces of the nanocatalyst. Therefore, the reaction could be accelerated in a hollow nanocatalyst due to the higher surface area. Reactions using the hollow nanocatalyst could be assisted by several factors. First, the confinement of reactants within the cavity could increase the steady-state concentration of the species in the rate-determining step of the reaction, and second, in some cases, the inner surface might not be as well capped as the outer surface and thus be more catalytically active. Therefore, the rate of the reaction increases due to the cage effect. Since the nanoreactor's wall thickness is small (a few nanometers), the electron can transfer across the wall during catalysis of electron transfer reactions. The surface-to-volume ratio of the hollow nanocatalyst is higher than any solid nanocatalysts because the surface area of the cavity adds to the nanocatalyst's outer surface area. This is another factor enhancing the catalytic properties of the hollow nanoparticles.[13] Although catalysis of nanoparticles has many advantages, some drawbacks are associated with the development of this type of nanocatalysts that must be addressed. The reshaping of nanocatalysts during catalytic reaction, products deposition on the nanocatalyst surface, and capping materials on the surface of colloidal nanocatalyst could decrease nanocatalyst stability, activity, limit their recyclability, affect their Fermi energy, and results in aggregation of nanocatalysts.

The galvanic replacement technique developed by Sun and Xia has allowed the synthesis of hollow metallic nanoparticles of various shapes.[5] Confinement of reacting materials within the hollow nanocatalysts has been demonstrated to enhance the catalytic efficiency of various chemical reactions.[13–18] This confinement is acquired either by designing the nanocatalyst to have a cavity, or by fixing the catalyst on the surface of the interior wall of an inert support. Nanocage catalysts made of a single metal (e.g., gold, platinum, or palladium) or two metals in a double shell arrangement have shown high catalytic efficiency due to the cage effect.[13–15] The idea of cage effect in catalysis was supported by comparing the catalytic activity of gold nanocages with that of the solid nanocatalysts of similar shapes (Figure 1.1).[19,20] To understand the mechanism of catalysis using hollow nanoparticles, two types of hollow nanocages with double shells, one with platinum around palladium and the other with palladium around platinum, and two single-shelled nanocages of platinum and palladium were synthesized.[14] The kinetic parameters (i.e., rate constant of the reaction, activation energy, entropy of activation, and frequency factor) of each double-shelled catalyst were comparable to those of the single-shelled nanocage of the same metal as the inside shell, indicating the reactions are taking place within the cavity. In addition, during catalysis, using double-shelled hollow nanoparticles with an inner surface made of gold plasmonic metal and a non-plasmonic the outer layer of platinum, as the reaction proceeded and the dielectric function of the interior gold cavity changed, the plasmonic band of the interior gold shell shifted. This strongly suggested that the reaction had taken place primarily in the cavity.[15] The cage effect was also observed for palladium nanotubes, which showed high catalytic efficiency for the Suzuki reaction.[17] Additionally, silver oxide prepared in situ on the interior

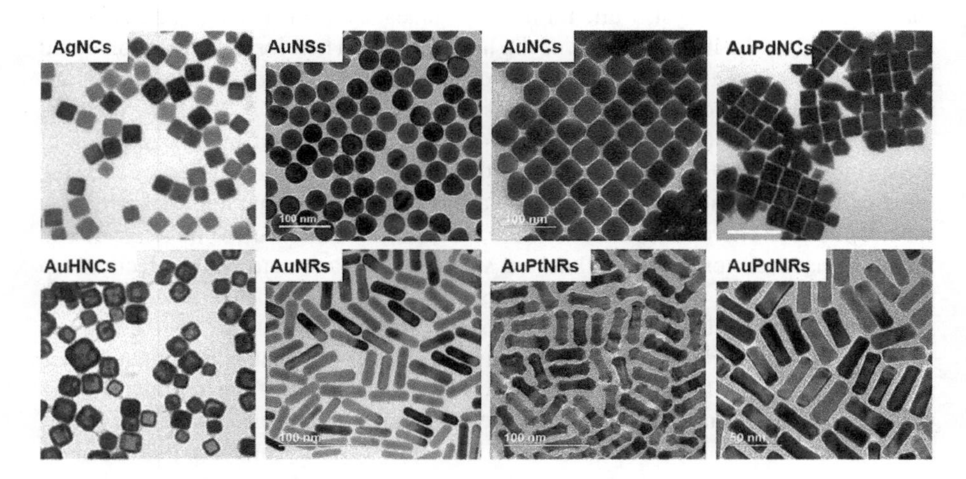

FIGURE 1.1 TEM images of single metal and bimetallic nanoparticles consist of Ag, Au, Pd, and/or Pt with various shapes. Silver nanocubes (AgNCs), gold nanospheres (AuNSs), gold nanocubes (AuNCs), gold-palladium nanocubes (AuPdNCs), hollow gold nanocages (AuHNCs), gold nanorods (AuNRs), gold-platinum nanorods (AuPtNRs), and gold-palladium nanorods (AuPdNRs) are among others presented here.

wall of gold nanocages showed high photocatalytic activity for the decomposition of an organic dye as the reactive radicals were confined within the gold nanocage catalysts.[21] Gold nanoparticles encapsulated in silica nanospheres catalyzed the production of hydrogen from formic acid.[22] Nanocatalysts fixed on the inner wall of porous metal-organic frameworks (MOFs) showed high catalytic activity due to the confinement of the reactant inside the voids present in the structure of the MOFs.[23] Polymer nanofibers, when used as a support for platinum and ruthenium nanocatalysts, showed high catalytic efficiency due to the cage effect.[24] Nickel, cobalt, iron, and their respective oxides encapsulated inside SiO_2 nanoshells effectively catalyzed reaction while maintaining thermal stability.[25] High electrocatalytic efficiency was also observed in hybrid inorganic nanostructures such as ruthenium and Cu_2S due to their electronic properties.[26] The kinetic parameters including the rate constant of the reaction, entropy of activation, and frequency factor for a nanocatalytic reaction are controlled by the following factors:

i) The concentration of the nanocatalyst and reacting materials.
ii) The number of successful collisions with the surface of the nanocatalyst in the rate-determining step.
iii) The electronic structure of the nanocatalyst, including the d-band center, which regulates the binding strength of adsorbates with the catalyst surface.
iv) The active surface area of the nanocatalyst and the reaction conditions (e.g., temperature and pressure).

Due to the small size of nanocatalysts, which in some cases approach the size of the reacting materials, the catalytic mechanisms are not clearly understood. Heterogeneous catalysis is possible when the reaction takes place on the surface of the nanocatalyst, while a homogeneous mechanism occurs when the catalysis proceeds through the formation of a complex between dissolved surface atoms in solution. Understanding these mechanisms would be greatly aided by determining which species were present on the nanoparticle surface during catalysis. Raman spectroscopy presents fingerprint vibrational spectra for molecules. Plasmonic nanoparticles can greatly enhance Raman signals through their enhancement of the electromagnetic fields of resonant light. This technique, known as surface-enhanced Raman spectroscopy (SERS), is sensitive down to the zeptomolar range (10^{-21}).[27] The high sensitivity of SERS has made it possible to use this technique to probe catalytic reaction as it occurs on the surface of the plasmonic nanocatalyst.[28,29]

1.2 AMMONIA AS A MEDIUM FOR INTERMITTENT RENEWABLE ENERGY STORAGE

Ammonia is one of the most widely produced commodities in the world, with 146 million tons being produced globally in 2015 with an estimated increase in production of 3–5% per year.[30,31] Ammonia is used in numerous applications, most notably as a vital agrochemical and as a precursor for pharmaceutical products.[32] In the United States, fertilizer manufacturing represented $18.2 billion in revenue in 2018.[33] Several trends are expected to increase demand for ammonia-based fertilizer in the short and long term. Crop production and fertilizer demand are positively correlated. In the short term, demand for crop production is expected to increase. In the longer term, the production of ammonia is expected to grow 65% by 2050, driven by population growth. The Agricultural Price Index (API) measures the money farmers earn on crop sales. The API was expected to increase in 2018; when income rises, farmers spend more on fertilizer. 80% of ammonia produced is used in fertilizer. Nitrogenous fertilizer comprises about 40% of the US Fertilizer industry, and mixed fertilizers account for another 25%. This equates to around $7.3 billion in ammonia consumption. Three large grain crops (wheat, rice, and corn) consume almost half the fertilizer used in agriculture. Approximately 45% of the total fertilizer purchased in the US is utilized in the corn industry, and 5.6 million tons of nitrogen fertilizer in 2010.[34] Therefore, the ammonia market for corn farmers should be at least $3.3 billion.

In addition to the compound's widespread utility, ammonia also holds great promise as a carbon-neutral liquid fuel for storing intermittent renewable energy sources when supply exceeds demand in the grid as well as for power generation due to the compound's high energy density and hydrogen content.[35,36] Ammonia can be used directly in alkaline fuel cells (AFCs) or indirectly as an H_2 source in proton exchange membrane fuel cells (PEMFCs). Ammonia can be a substitute for H_2 as a combustion fuel with superior advantages in terms of energy density, ease of liquefaction, and high hydrogen content (for example, the hydrogen content in liquid ammonia is 17.6% by weight, compared with 12.5% for methanol). The volumetric energy density of ammonia is 13.6 GJ m^{-3} (10 atm at 298 K), which is compared with 5.3 GJ m^{-3} for hydrogen at 700 bar pressure.[37] Thus, sustainable nitrogen fixation lies at the nexus of food and energy chemistry (Figure 1.2). Conventional ammonia synthesis mainly relies on the Haber–Bosch process, which converts N_2 and H_2 to NH_3 at high operating pressures (150–250 bar) and temperatures (350–550 °C) over iron-based catalysts.[30] The extreme condition requirements for this process necessitate high-cost demands for centralized infrastructure that should be coupled with the global distribution system. Additionally, this process consumes 3–5% of the global natural gas supply, and 60% of global hydrogen production, and emits 450 million metric tons of CO_2 annually.[38]

Various technologies could be developed to store surplus renewable electricity when supply exceeds demand in the grid in the form of liquid ammonia. The dispatchable and transportable ammonia fuel can then be utilized for different energy and agriculture sectors.

Although ammonia has been utilized in vehicles for many decades, the development of ammonia as an energy carrier was primarily due to its use as a hydrogen carrier. Given ammonia's potential for widespread use as an energy carrier or liquid fuel, many direct end-use applications are envisioned for commercialization. To use ammonia as a fuel, it is essential to compare the cost of ammonia with other liquid fuels such as diesel, LPG, gasoline, and marine fuel bunker (Table 1.1). The international price of ammonia has ranged between $250 and $400 per ton of ammonia depending upon the input feedstock (e.g., natural gas or coal) with an average price of $300 ton^{-1} NH_3, including the transportation costs. As presented in Table 1.1, ammonia fuel can be competitive with current carbon-based fuels, assuming the end-use technologies have similar energy efficiency[39]. It is important to note that if the

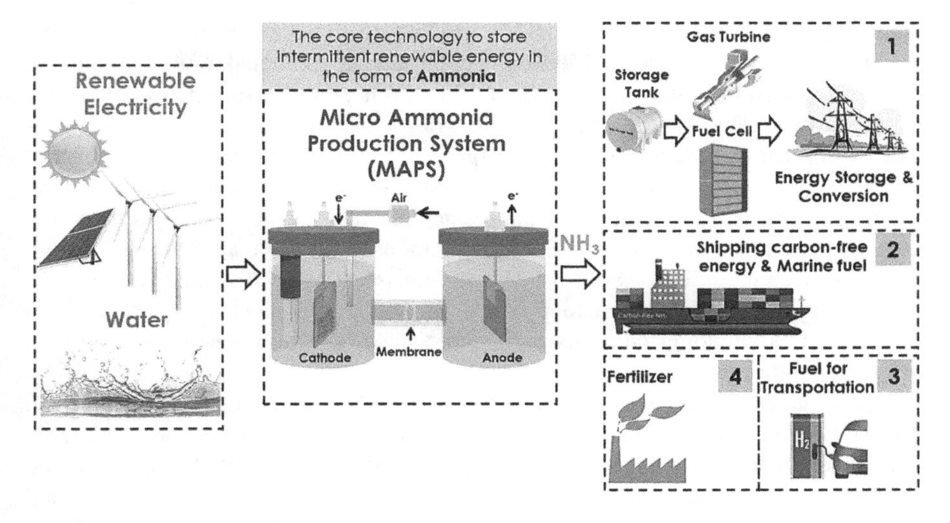

FIGURE 1.2 Schematic diagram of photo-electrochemical systems for value-added chemicals production with various applications in energy, transportation, and agriculture sectors.

carbon tax is included in the current prices (~ $30 per ton of carbon dioxide), then green routes for making synthetic fuels can be even more appealing.

Internal combustion engines (ICE) can use ammonia as an alternative fuel. For instance, a spark-ignition engine has been tested in South Korea that runs on a hybrid fuel containing 70% ammonia and 30% gasoline.[40] The use of ammonia fuel has also been expanded to compression ignition engines. However, the NO_x emission resulting from ammonia fuel combustion should be controlled and mitigated by selective catalytic reduction technologies. Ammonia is also an attractive marine fuel for the shipping industry, which promises to lower the sulfur content of fuels recently regulated by the International Maritime Organization. A recent study demonstrated that ammonia-fueled marine ICEs have comparable performance and emission

TABLE 1.1
Relative Properties and Costs of Ammonia Compared with Liquid Fossil Fuels

Fuel	P (bar)	Density (kg m^{-3}) (15 °C)	LHV (kWh kg^{-1}) (25 °C)	LHV (MWh m^{-3}) (25 °C)	Cost (USD kg^{-1})	Cost (USD kWh^{-1})
Ammonia	10	603	5.18	3.12	0.3	0.058
Diesel	1	846	12.1	10.2	1.00 (USA)	0.083
LPG	14	388	12.6	4.89	1.00 (Germany)	0.079
Gasoline	1	736	12.1	8.87	1.81 (Japan)	0.15
Bunker Fuel	1	980	10.8	10.6	0.59 (Global average)	0.055

metrics in power generation and NO_x emissions to those using diesel fuel.[41] The European-based Transport and Environment Group has estimated to use ammonia in marine applications with the capacity of 1200 TWh year^{-1} in Europe by 2050[42] for the current EU electricity generation in 2015 of 3200 TWh year^{-1}. Additionally, generators that can produce and store ammonia can be used as a substitute for diesel generators in remote locations for power generation to complement solar and wind electricity. The use of ammonia as a supplementary fuel in coal- and gas-fired power generators and gas turbines have also been demonstrated in Japan. This would help transition such facilities toward lower carbon emissions as soon as competitive technologies are being developed to produce ammonia at competitive prices.[43] This would lead to ammonia being used for large-scale renewable energy storage and power generation at the grid level.

Ammonia can be utilized directly in ammonia fuel cells or indirectly in hydrogen fuel cells. For example, ammonia cracking reactors can produce pure hydrogen, which then hydrogen feeds the fuel cell. High temperature (600–900 °C) solid oxide fuel cells (SOFCs) can use ammonia to generate electricity without any external reformer or cracking reactors.[44] In SOFCs, ammonia decomposes to nitrogen and hydrogen at high temperatures. The low-temperature alkaline fuel cell is another type of direct ammonia fuel cell (DAFC) that can use ammonia with the following anodic and cathodic reactions:

$$\text{ammonia oxidation}: \quad 4NH_3 + 12OH^- \rightarrow 2N_2 + 12H_2O + 4e^- \tag{1.1}$$

$$\text{oxygen reduction}: \quad O_2 + 2H_2O + 4e^- \rightarrow 4OH^- \tag{1.2}$$

In both types of direct ammonia fuel cells (SOFCs and DAFCs), there is a potential for NO_x generation; however, in these early-stage technologies, the control of NO_x emissions should be thoroughly investigated.

1.3 GLOBAL NITROGEN CYCLE

A transition from an economy dependent on fossil fuels to one based on ammonia must consider the environmental impacts caused by the production and use of this new fuel. Similar to biogeochemical processes contributing to the global carbon cycle, the planetary nitrogen cycle is complex and not entirely well understood. Technologies that are being developed for artificial nitrogen fixation processes with an enormous impact on lowering CO_2 emissions should not create another problem of releasing excess NH_3 and NO_x emissions to the environment. Atmospheric, land-based, and marine cycles are major pathways for exchanging N_2 and the various reactive forms of nitrogen (Figure 1.3). The anthropogenic activities for nitrogen fixation (e.g., Haber–Bosch process) are shown on the left. The upward arrows show re-emissions to the atmosphere by various processes. This diagram indicates that anthropogenic nitrogen fixation is equivalent to the natural processes. Ultimately, the Earth should manage the cycling and re-emission of extra fixed nitrogen, which has been doubled since the invention of the Haber–Bosch process over a century ago. A significant portion of this fixed nitrogen is in the form of "nitrate," including

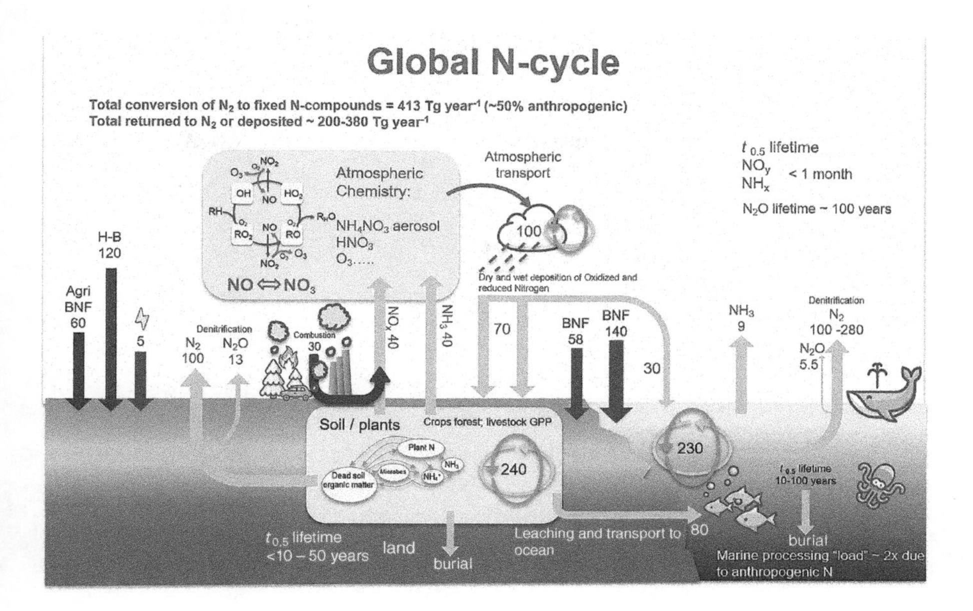

FIGURE 1.3 The planetary nitrogen cycle. Planetary nitrogen cycles and fluxes in Tg per year (BNF, biological nitrogen fixation; H-B, Haber–Bosch nitrogen fixation; GPP, gross primary productivity). Curved "combustion" arrow shows nitrogen fixation from fossil fuel and other combustion process and resultant emissions of NO_x. compounds to the atmosphere. 1 Tg = 1 teragram = 1 × 10^6 tonne. Reprinted with permission from MacFarlane et al.[39] Copyright 2020, Elsevier.

surface and groundwater nitrate (NO_3^-) pollution, eutrophication of freshwater systems, and massive killing of aquatic organisms in coastal regions that comprise so-called "dead zones" due to depleted oxygen.[45,46] It is vital to thoroughly investigate and understand the global nitrogen cycles while studying and developing ammonia synthesis technologies.

2 Conventional Methods for Nitrogen Fixation

2.1 INTRODUCTION

The nitrogen cycle is one of the most important biogeochemical cycles on Earth because nitrogen is a crucial nutrient for all life forms, including bacteria, plants, and humans. Nitrogen is an essential component of nucleic acids, amino acids, vitamins, cofactors, and hormones. In nature, various nitrogen-containing species are interconverted on a scale of millions of metric tons per year.[45] As nitrogen is a major fertilizer component in agriculture, it plays a critical role in food production, responsible for sustaining a growing global population.

The atmosphere comprises 78% gaseous nitrogen (N_2), a majority of Earth's available nitrogen. N_2 must be fixed to its reactive forms (e.g., ammonia), so that it can be used by plants and microorganisms. Microorganisms that live in association with leguminous plants are responsible for the 120 million metric tons of nitrogen fixed through natural processes and become available to the biosphere.[30] Nitrogenase, the enzyme responsible for catalyzing the six electrons N_2 reduction reaction to NH_3, has been the focus of nitrogen cycle research for a few decades.[47–49] Artificial nitrogen fixation has complemented biological nitrogen fixation through the Haber–Bosch process, which fixes the same amount of N_2, nature does each year.[50] The industrial Haber–Bosch process is responsible for feeding nearly half of the world's population and will become more vital as the population continues to grow.[51]

2.2 NATURAL NITROGEN FIXATION

Biological nitrogen fixation occurs naturally in diazotrophic microorganisms through the enzyme nitrogenase. Nitrogenase operates at mild conditions (<40 °C, atmospheric pressure) compared with heterogeneous catalysts used in the Haber–Bosch process (300–500 °C, >150 bar). The study of this enzyme is of great importance to meet the grand challenge of sustainable and efficient ammonia synthesis. Ammonia synthesis from dinitrogen by nitrogenase takes place according to the following equation under optimal conditions where ATP is adenosine triphosphate, ADP is adenosine diphosphate, and Pi is inorganic phosphate:[52]

$$N_2 + 8H^+ + 16MgATP + 8e^- \rightarrow 2NH_3 + H_2 + 16MgADP + 16Pi \quad (2.1)$$

The reaction involves the obligatory hydrolysis of ATP to release stored chemical energy and thermodynamic or kinetic barriers of nitrogen reduction. For every molecule of nitrogen reduced, two molecules of ammonia are generated, and protons are also reduced to form one molecule of hydrogen. Therefore, 25% of the energy consumed results in hydrogen production. In the absence of nitrogen or other substrates, nitrogenase promotes the reduction of protons.

Three distinct nitrogenase enzymes with similar characteristics have been found where they are distinguished by their metal-centered catalytic cofactors (co): FeMo-co, FeFe-co, and VFe-co. The most widely studied and understood nitrogenase enzyme contains the FeMo-co, and is known as MoFe nitrogenase. X-ray crystallography has given insight into the crystal structure of the MoFe nitrogenase enzyme, and a representation of the structure is shown in Figure 2.1a. Nitrogenase consists of two multi-subunit proteins, both oxygen-sensitive, with one protein serving as a catalytic domain and the other serving as a reducing domain. The MoFe protein is the catalytic protein to which nitrogen binds and is reduced to ammonia. In contrast, the second protein, or the Fe protein, hydrolyzes MgATP molecules and transfers electrons to the MoFe protein for catalysis. The basic components of these proteins and their functions are presented in Table 2.1. The Fe protein is a homodimer, and each subunit in the protein contains a nucleotide binding site for a MgATP molecule and two cysteine residues to which the bridging [$4Fe_4S$] cluster binds. The MoFe protein is a larger $\alpha_2\beta_2$ tetramer containing two Fe_8S_7 'P clusters' (one bridging cluster between each α and β subunit dimer) and two FeMo cofactors (Fe_7MoS_9C, located in the α subunit). The P cluster plays an exclusive role in transferring electrons originating from coupled ATP hydrolysis in the Fe protein to the FeMo-co of the MoFe protein. The electron transfer to each component is shown in Figure 2.1a.

The complete molecular structure of nitrogenase, specifically FeMo-co, has been determined by identifying an interstitial carbon atom at the center (Figure 2.1b).[53] Despite this progress, understanding of the electronic structure of the FeMo-co is still lacking. Electronic studies suggest that FeMo-co only cycles through one redox couple, with one resting stage M^N and a one electron-reduced stage M^R[54]. Although

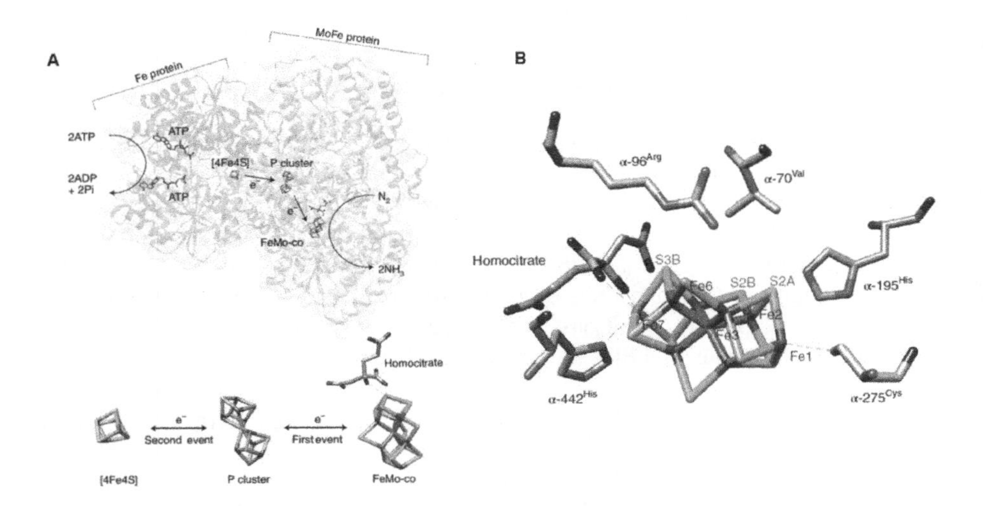

FIGURE 2.1 Nitrogenase enzyme structure and functions. (a) Diagram of one half of the nitrogenase complex and electron transfer. (b) Detailed diagram of the FeMo cofactor and the surrounding environment. Reprinted with permission from Foster et al.[52]

TABLE 2.1
Basic Components and Functions of the Nitrogenase Enzyme

Protein	Domain	Function
Fe protein homodimer (~66 kDa)	Fe_4S_4, F cluster	Facilitates hydrolysis of MgATP and electron transfer to the MoFe protein
	Nucleotide binding sites	Facilitates binding of MgATP
MoFe protein $\alpha_2\beta_2$ tetramer (~240 kDa)	FeMo-co clusters	Catalyzes reduction of nitrogen to ammonia, buried to prevent access to H_2O and improve nitrogen selectivity over hydrogen evolution
	Fe_8S_7, P clusters	Responsible for transferring electrons to the FeMo-co from the Fe_4S_4 cluster of the Fe protein

Reprinted with permission from Foster et al.[52]

Mo^{iv} was known to be the oxidation state of the molybdenum, recent studies propose the reassignment of the oxidation state to Mo^{iii}.[55] This finding was further complemented by other studies that assign the charges of the iron and molybdenum atoms in the FeMo-co so that substrate reactivity with the Fe and Mo atoms in the FeMo-co can be better understood.[56] Studies have shown that three of the seven iron atoms in the FeMo-co (iron atoms labeled Fe_1, Fe_3, and Fe_7 in Figure 2.1b) are relatively reduced compared with the remaining four iron atoms. Given this information, a FeMo-co with three iron atoms in the 2^+ oxidation state, four iron atoms in the 4^+ oxidation state, and a molybdenum atom with a 3^+ oxidation state, are consistent with these observations of the FeMo-co.

With the existence of three types of metal-centered catalytic cofactors (FeMo-co, FeFe-co and VFe-co), the specific substrate binding site is still not well understood. Concerning the FeMo-co, the involvement of amino acids in the catalytic reaction and the spatial positioning (Figure 2.1b) suggests the active site to be the Fe-S face, but determining the location and binding mode of dinitrogen within the nitrogenase remains a challenge. Despite knowledge of the nitrogenase structure and active site, the mechanism of nitrogenase-mediated dinitrogen reduction to ammonia remains unsolved. Amino acid substitutions and freeze-quench trapping, however, have isolated intermediates in support of draft mechanisms. Lowe and Thorneley developed an eight-step kinetic model to reduce dinitrogen to ammonia by nitrogenase (Figure 2.2). Hoffman et al. proposed a nitrogenase mechanism of N_2 activation and reduction in agreement with the Lowe–Thorneley kinetic model.[48,54] In the Lowe–Thorneley kinetic model, one proton and one electron bind to the cofactor during each stage. This draft mechanism proposes H_2 generation to proceed through a reductive elimination of hydrides producing a highly reduced FeMo-co intermediate stage (E_4). At stage E_4, four protons are bound to the FeMo-co, including two protons that are bound to two iron atoms each (Fe–H–Fe). At this stage, N_2 can bind to the FeMo-co, and the dinitrogen triple bond can be split.

Multiple computational studies have also proposed possible mechanisms for dinitrogen reduction. Varley et al.[57] developed a model that uses density functional theory

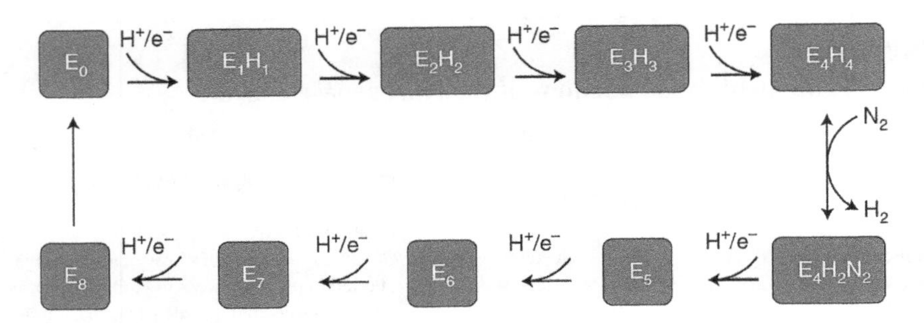

FIGURE 2.2 Single electron-proton transfer model for nitrogenase-mediated nitrogen fixation. Simplified representation of the Lowe–Thorneley kinetic scheme for dinitrogen reduction.[48] The full mechanistic model also features N_2 binding at E_3. Reprinted with permission from Foster et al.[52]

(DFT) calculations to explain how belt-sulfur atoms dissociate as H_2S (S_2B sulfur, Figure 2.1b) from the FeMo-co reveal the reactive Fe sites. The work was based on structural experiments supporting the displacement of a belt-sulfur atom by CO.[58] According to the model, this step is critical to the initiation of the N_2 reduction process. The subsequent re-sealing of the active Fe site by H_2S re-absorption, required to desorb the second NH_3 product, releases H_2, which can account for the requisite H_2 produced per reduced N_2 by nitrogenase in the Lowe–Thorneley kinetic model.[58] Mechanisms determined in additional studies highlight the vital role the interstitial carbon and protein environment play in initiating the dinitrogen reduction process.[59]

Studying the nitrogenase structure and developing molecular catalysts has allowed a deeper understanding of the structural environment for efficient N_2 reduction to NH_3. The nitrogenase structure has primarily utilized amino acid substitutions to control the enzyme catalytic environment, allowing the study of the substrate and the intermediates for ammonia production. Homogeneous catalysis has paved the way to develop and understand ligand systems and molecular catalyst structures for ammonia synthesis. These findings from the nitrogenous and molecular catalysts must be supplemented with mechanistic insights to design heterogeneous catalysts for nitrogen reduction reaction rationally.

2.3 THERMOCATALYTIC PROCESS

In 1909, Fritz Haber and Carl Bosch developed an artificial nitrogen fixation process called the Haber–Bosch process. This process enabled the production of ammonia and the transformation of our society and lives through this revolutionary invention. Since then, ammonia has been used extensively in making fertilizers, which is responsible for increasing the population from two to more than seven billion people during the last century. The global production of ammonia is approximately 150 million metric tons and is projected to increase between 2–3% per year.[60] The use of ammonia as long-term energy storage (e.g., months) can be considered an attractive alternative to short-term storage (e.g., batteries). By storing renewable energy in the ammonia chemical bonds, energy can be delivered to end-users by on-demand hydrogen production from ammonia (17.6 wt% hydrogen) combined with fuel cells.

Despite an enormous impact and potential of ammonia as fuel and fertilizer, production of ammonia through the Haber–Bosch process (>96% of ammonia is currently produced through this route) using fossil fuels as feedstock (e.g., natural gas, coal, oil) results in negative impacts on sustainability and carbon-free society. The Haber–Bosch process is currently one of the largest global energy consumers (>1% global energy consumption) and greenhouse gas emitters (1.2% of the global CO_2 emission). It is important to note that the optimization of the process has been conducted, assuming fossil fuels as the only viable energy source. Therefore, most efforts have focused on improving hydrogen production efficiency through the steam methane reforming (SMR) process ($CH_4 + H_2O \leftrightarrow CO + 3H_2, \Delta H^o_{298K} = 206.2\, kJ\, mol^{-1}$), and the process is not optimized to reduce carbon emissions. There are technical challenges that must be addressed to develop and deploy a new chemical process for ammonia synthesis with increased efficiencies and a remarkable decrease in CO_2 emissions. For example, if i) the Haber–Bosch process is decoupled from the SMR process, ii) electric compressors replace condensing steam turbine compressors, and iii) alternative ammonia separation techniques are utilized to decrease the operating pressure, then a new process allows the adoption of ammonia synthesis process for small-scale production, which is in agreement with the intermittency and isolation of renewable energy sources (e.g., solar and wind). This will also help the widespread utilization of ammonia in the energy storage sector.

A modern ammonia synthesis based on the Haber–Bosch process can be separated into two steps (Figure 2.3a):[61] the first step is hydrogen production during the SMR process. The second process is ammonia synthesis by the Haber–Bosch reactor. Hydrogen is produced by primary and secondary SMR reactors, followed by a two-stage water-gas shift reactor, CO_2 removal, and methanation. The first SMR reactor operates at around 850–900 °C and 25–35 bar, and the combustion of methane fuel provides the energy required for this endothermic process through furnace tubes that run through the catalyst bed. In the second SMR reactor, the air is compressed and fed to the reactor to supply the heat of reaction at 900–1000 °C. The SMR outlet mixture of carbon monoxide (CO), hydrogen (H_2), steam, and methane are introduced into the two-stage water-gas shift (WGS) reactor to enhance the production of H_2 according to Equation 2.2:

$$CO + H_2O \rightarrow CO_2 + H_2 \qquad \Delta H^o_{298K} = -41.2\, kJ\, mol^{-1} \tag{2.2}$$

A methanation reaction occurs at the end of the process to convert any remaining CO to methane to prevent the Haber–Bosch catalyst poisoning. Overall the SMR process is endothermic, but the high-temperature requirement for the reaction and the cooling need for the WGS reaction results in substantial heat loss, which is further used to generate high-pressure steam in turbines. The ammonia production stage consists of the Haber–Bosch reactor where hydrogen and nitrogen react at 150–250 bar and 400–500 °C using an iron-based catalyst according to Equation 2.3:

$$\frac{1}{2}N_2 + \frac{3}{2}H_2 \leftrightarrow NH_3 \tag{2.3}$$

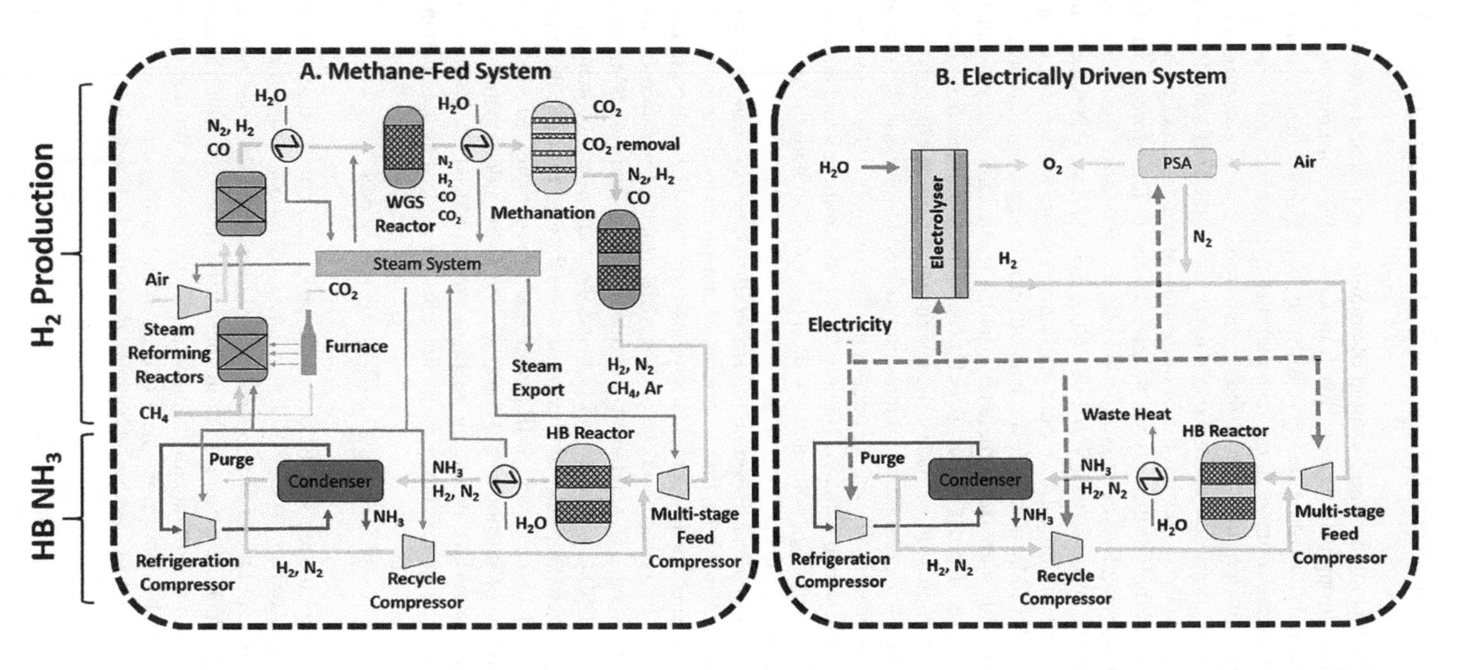

FIGURE 2.3 Schematic diagram of (a) a typical conventional methane-fed Haber–Bosch process and (b) an electrically powered alternative. Hydrogen and ammonia production stages are separated for illustration purposes to identify similitudes and differences between both technologies. Yellow lines are process gas, dark blue lines are water/steam, light blue lines are air, purple lines are ammonia, and dashed lines are electricity. Reprinted with permission from Smith et al.[61]

This reaction is exothermic ($\Delta H^o_{298K} = -45.9\,kJ\,mol^{-1}, \Delta G^o_{298K} = -16.4\,kJ\,mol^{-1}$, $K_{eq} = 750$), and there is a too high kinetic barrier to break the triple bond of N≡N, and high temperatures are required to break this bond at appropriate rates. As ammonia decomposes to N_2 and H_2 at temperatures greater than 400 °C at 1 bar, the Haber–Bosch reaction is operated at high pressures (15–40 MPa) to shift the reaction equilibrium toward the formation of ammonia. Low equilibrium single-pass conversion (~15%) necessitates the use of multi-pass reactors. In each stage, ammonia product is removed by condensation, and the inert gases (methane and argon) are purged and recycled to the SMR furnace.

Electrification of the chemical industry necessitates the use of renewable electricity for chemical manufacturing.[62] In the Haber–Bosch process, hydrogen can be produced by the electrolysis of water and converted to ammonia using a conventional Haber–Bosch reactor (Figure 2.3b). Nitrogen (N_2) is separated from the air through pressure swing adsorption, suitable for small-scale applications.

The Haber–Bosch process has been improved and optimized continuously since its invention over a century ago; however, the current catalyst is very similar to the original iron (Fe) catalyst.[63] Studies have demonstrated Fe (111) single crystal structure is the most reactive crystal face for ammonia synthesis due to the highly coordinated Fe sites, which experience the most considerable electronic fluctuations.[64] The addition of promoters (e.g., Al_2O_3, K_2O) can improve the catalyst performance in the high-temperature and high-pressure environment of the Haber–Bosch reactor. In addition, these promoter oxides can increase the Brunauer–Emmett–Teller surface area of the hybrid catalyst.[65] Ozaki et al. proposed the energetics of chemical adsorption and desorption of nitrogen species on the catalyst surface could follow the so-called "volcano" plots.[66] Ruthenium-based catalysts have also received attention due to their optimum energy on the volcano graph, leading to the commercialization of the first ammonia synthesis reactor using non-ferrous catalysts.[67] Ru-based catalysts have a higher cost and lower durability than Fe-based catalysts,[68,69] and therefore, research and development should be focused on exploring potential catalyst supports, such as boron nitride,[70] $BaCeO_3$ nanocrystals,[71] lanthanide oxides (MgO),[72] CeO_2,[73] graphitic nanofilaments,[74] and zeolites.[75] Recent work demonstrated that a calcium aluminum oxide-based electride ($[Ca_{24}Al_{28}O_{64}]^{4+}$ $(e^-)_4$) acts as electron-donating support for a Ru catalyst, resulting in ammonia production rates as high as 2120 μmol g^{-1} h^{-1} (a gas inlet of H_2/N_2 (3:1), 1 atm, 673 K), compared with traditional alumina or calcium oxide-supported Ru catalysts with ammonia production rates of 50–160 μmol g^{-1} h^{-1}.[76]

Currently, the state-of-the-art methane-fed Haber–Bosch reactor emits ~ 1.6 ton of carbon dioxide per ton of ammonia production. This value of emission will further increase if the extraction and transport of natural gas are included. The majority of CO_2 emissions (~75% or 1.2 t_{CO2} t_{NH3}^{-1}) originates from hydrogen production through the SMR process.[77] The remaining ~ 25% is consumed to provide the required energy for an endothermic reaction (Figure 2.4). In the electrically driven Haber–Bosch process, the CO_2 emissions will decrease considerably from 1.6 t_{CO2} t_{NH3}^{-1} to 0.38 t_{CO2} t_{NH3}^{-1} if methane is replaced with renewable energy sources.[61] For example, assuming that the renewable-powered Haber–Bosch process requires a 38.2 GJ t_{NH3},

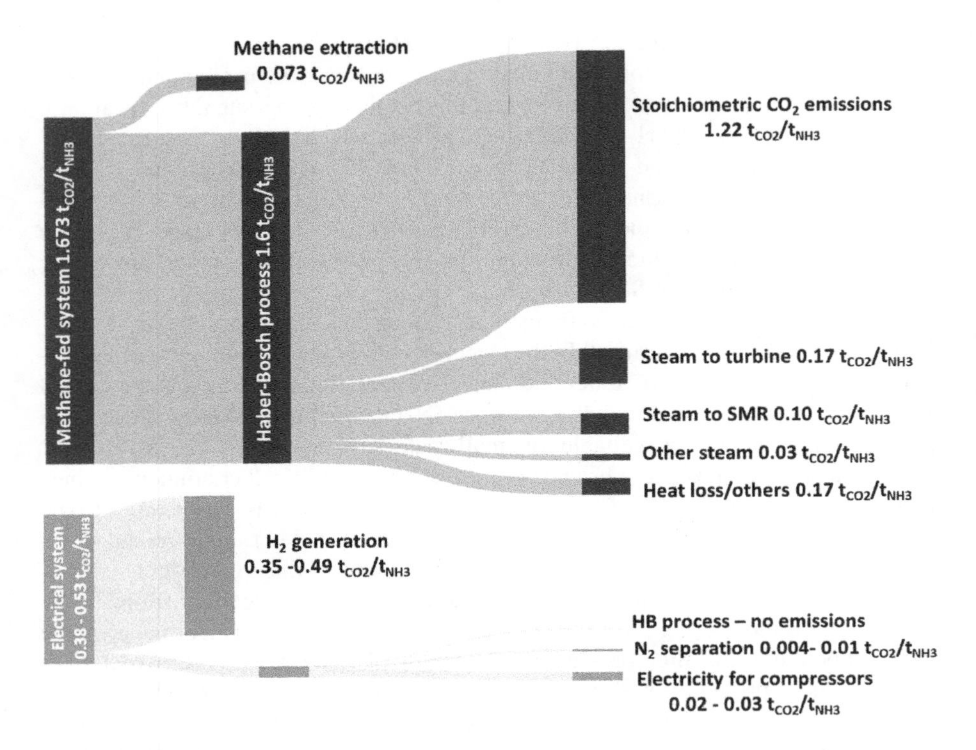

FIGURE 2.4 Sankey drawing comparing the attributions of direct CO_2-eq emissions arising from the methane-fed and the electrically driven Haber–Bosch processes (range of values depending on the size of wind turbines). The stoichiometric CO_2 emissions are shown to highlight the minimum level of direct CO_2 emissions achieved by the methane-fed system without carbon capture. The additional CO_2 emissions are allocated proportionally to the significant energy consumers. Reprinted with permission from Smith et al.[61]

35.5 GJ t_{NH3} is consumed for hydrogen production with 60% efficient electrolyzer and ~2.7 GJ t_{NH3} is used for the N_2 separation and the Haber–Bosch loop compressors. Therefore, a wind-powered ammonia synthesis reactor will have a carbon intensity of 0.12–0.53 t_{CO2} t_{NH3}^{-1}.[61] This range is calculated based upon the size of wind turbines and average wind speed. The estimated 2018 global production of renewable wind and solar energy of 2480 TWh is sufficient to meet the global demand for ammonia, which was ~ 150 million metric tons (requiring 1670 TWh electricity) in 2015.[78,79]

3 Electrocatalytic Nitrogen Reduction Reaction for Ammonia Synthesis

3.1 INTRODUCTION

Developing sustainable and environmentally friendly ammonia production methods that consume significantly less energy than the current industrial process is imperative. As the cost of renewably derived (e.g., solar and wind) electricity continues to decrease given the rapid progress in technology and economies of scale, there is a growing interest in NH_3 electrosynthesis from N_2 and H_2O under ambient conditions in an electrochemical cell.[80] Electrocatalytic fixation of nitrogen is a form of artificial synthesis that mimics the natural nitrogen enzymatic process.[53] This approach can provide an alternative pathway to the Haber–Bosch process for clean, sustainable, and distributed ammonia synthesis as well as the storage of surplus renewable energy in the form of NH_3 fuel at times of excess supply in the grid.[81,82] Electrification of ammonia synthesis on a large scale requires an effective electrocatalyst that converts N_2 to NH_3 with a high yield and Faradaic efficiency (FE). An efficient catalyst should bond atoms and molecules with an intermediate strength so that catalysis will not be limited by weak adsorption of reactants or strong desorption of products. To date, most studies have shown low electrocatalytic activity and selectivity for NH_3 production mainly due to the high energy required for N≡N cleavage and due to the competition with the hydrogen evolution reaction (HER).[83–87] Various strategies for the photo- and electrocatalytic nitrogen reduction reaction (NRR) have been proposed in the literature, aiming to design novel electrodes, electrolytes, and photo-electrochemical cells to enhance the rate of ammonia synthesis (Figure 3.1). Additionally, the operating temperature and electrode reactions of a photo-electrochemical cell are determined by the type of the electrolyte. In the ambient pressure and temperature conditions, aqueous, non-aqueous, and polymer electrolyte membranes have been utilized as the electrolyte.[88] In the intermediate temperature range, molten chlorides,[89] molten hydroxides,[90] and solid acids[91] are the most commonly used electrolytes. In the high-temperature region, carbonates[92] and solid-state proton or oxide-ion conductors with perovskite, fluorite, or pyrochlore structure[93] were evaluated as the electrolyte. Protons (H^+), hydroxide ions (OH^-), oxide ions (O^{2-}), and nitride ions (N^{3-}) are typically mobile charge carriers in these electrolytes. The electrode reactions will vary depending on the type of charge carrier and the feed gas composition (Figure 3.2).[94] The selectivity of N_2 molecules on the surface of nanocatalysts over the kinetically favored HER has been demonstrated to be one of the major challenges in enhancing the rate of

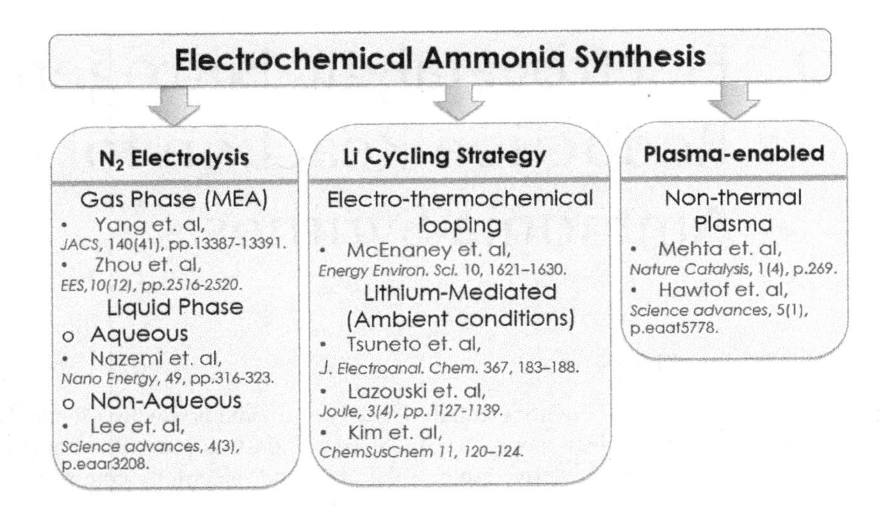

FIGURE 3.1 Background of various electrochemical approaches for nitrogen fixation in the literature. There are three main strategies for electrochemical nitrogen fixation: 1) N_2 electrolysis, 2) Li cycling strategy, and 3) Plasma-enabled nitrogen fixation.

electrochemical NRR in aqueous solution under ambient conditions.[95–97] In addition, achieving a high ammonia yield rate is vital to make this process industrially feasible. A recent report suggested an ammonia formation rate of 10^{-6} mol cm^{-2} s^{-1} and a FE of ~50% would be required for commercial applications.[98] Achieving high production energy efficiency (maintaining low overpotentials for the anodic and cathodic half-reactions) is another critical factor that needs to be addressed.[99] It is important to note that while plasma electrolytic systems result in remarkably high N_2 selectivity and ammonia formation rate, it suffers from a very low energy conversion efficiency as a result of a high voltage input (i.e., 500 V).[100]

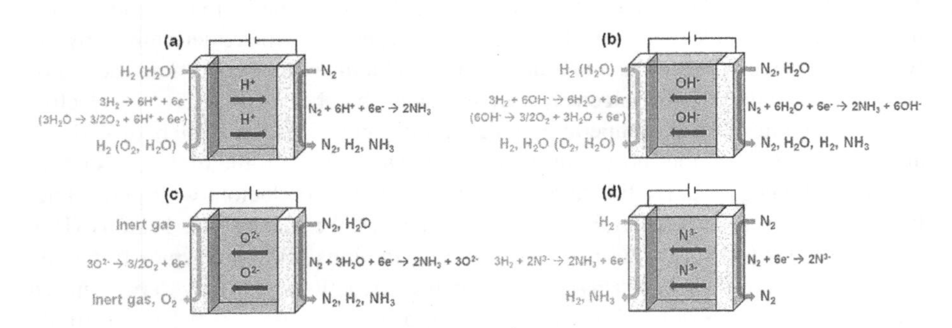

FIGURE 3.2 Electrode reactions for ammonia synthesis in electrochemical cells using (a) proton (H^+), (b) hydroxide ion (OH^-), (c) oxide ion (O^{2-}), and (d) nitride ion (N^{3-}) conducting electrolytes. Reprinted with permission from Refs.[94,101]

3.2 FUNDAMENTAL UNDERSTANDING OF ELECTROCATALYTIC N_2 REDUCTION

The N_2 molecule stability and the high dissociation energy of the N≡N triple bond (941 kJ mol^{-1}) make this molecule an appropriate candidate for providing an inert atmosphere (e.g., glovebox). Even though the dissociation energy of the N≡N triple bond is lower than those of acetylene (HC≡CH, 962 kJ mol^{-1}) and carbon monoxide (C≡O, 1070 kJ mol^{-1}), the cleavage of the first bond in N≡N requires almost two times more energy (410 kJ mol^{-1}) than those in acetylene and carbon monoxide.[88,101] This indicates that the initiation of dissociating the triple bond of N_2 is a challenging task. Lack of a permanent dipole, a low proton affinity (493.8 kJ mol^{-1}), a negative electron affinity (−1.90 eV), high ionization potential (15.84 eV), and its large energy gap (10.82 eV) between the highest occupied and lowest unoccupied molecular orbitals hinder electron transfer reactions.[102] The hydrogenation of N_2 to NH_3 is thermodynamically favored; however, the energies of possible intermediates (e.g., diazene (N_2H_2) and hydrazine (N_2H_4)) are so high that the hydrogenation of N_2 does not occur spontaneously.[103]

Electrochemical reduction of N_2 to ammonia is a multistep process involving six electrons and protons. The half-reactions of N_2 reduction and water oxidation and their corresponding equilibrium potentials at 25 °C are as follows:

$$N_2(g) + 6H^+(aq) + 6e^- \leftrightarrow 2NH_3(aq) \quad E^o = 0.058\ V\ vs.\ RHE \tag{3.1}$$

$$3H_2O(l) \leftrightarrow \frac{3}{2}O_2(g) + 6H^+(aq) + 6e^- \quad E^o = 1.23\,V\ vs.RHE \tag{3.2}$$

The overall reaction is:

$$N_2(g) + 3H_2O(l) \leftrightarrow 2NH_3(aq) + \frac{3}{2}O_2 \quad E^o = 1.17\,V\ vs.RHE \tag{3.3}$$

Theoretically, when a sufficiently negative potential is applied to an electrode, it should drive the reduction of N_2 at room temperature and atmospheric pressure. However, the equilibrium potential of N_2 reduction is similar to the hydrogen evolution reaction (HER) according to the following reaction:

$$H_2(g) \leftrightarrow 2H^+(aq) + 2e^- \quad E^o = 0.00\ V\ vs.\ RHE \tag{3.4}$$

The HER only requires two electrons and therefore is kinetically more facile than six-electron NRR. In addition to the thermodynamic factors, catalytic NRR involves three steps:

1) adsorption of N_2 on the active sites of the catalyst,
2) the protonation of adsorbed N_2 and N≡N triple cleavage to produce ammonia, and
3) desorption of ammonia from the catalyst surface.

Depending on the adsorption and protonation modes of N_2, the mechanisms are classified into associative and dissociative pathways (Figure 3.3).[88,104,105] In the

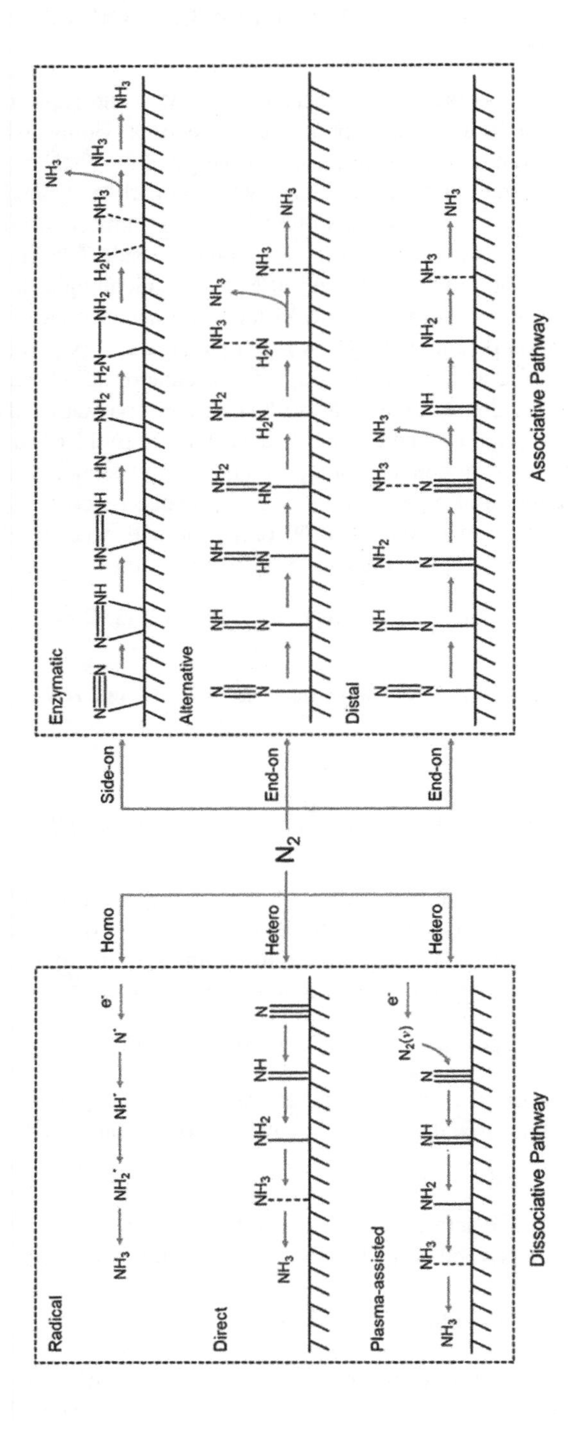

FIGURE 3.3 Proton-coupled electron transfer pathways for ammonia synthesis: (left) dissociative pathways involving radical, direct, and plasma-assisted processes and (right) associative pathways involving enzymatic, alternative, and distal processes. Reprinted with permission from Ref.[104]

dissociative pathways, the N≡N triple bond is broken before protonation. The dissociation of the N_2 molecule can occur directly on specific catalysts under thermo-catalytic conditions or through the plasma-assisted process to produce vibrationally excited nitrogen molecules ($N_2(\upsilon)$), forming the surface adsorbed nitrogen atoms. Homogeneous radical chemistry is categorized as a dissociative pathway, where N radicals (N^*) formed in the plasma will combine with hydrogen radicals in the gas phase to form ammonia. In the associative pathways, N_2 molecules are adsorbed onto the catalyst surface in "end on" and "side on" geometries. In the "side on" mode as exemplified by nitrogenase in enzymatic N_2 fixation, both atoms of N_2 bond with the catalyst surface, followed by the alternating hydrogenation of each nitrogen, with N≡N triple bond cleavage forming two ammonia molecules. In the "end on" mode, only one atom of the N_2 molecule bonds with the catalyst surface. Depending upon the protonation steps, the "end on" mode can be further categorized into alternating and distal pathways. In the alternating associative pathway, the nitrogen atoms of adsorbed N_2 are protonated alternatively. In the distal associative pathway, the terminal nitrogen (i.e., the N not bound to the catalyst) is protonated preferentially, with N≡N bond cleavage to release one ammonia molecule and to leave one N atom adsorbed on the catalyst surface for further protonation to NH_3.[104]

To highlight the critical role of selectivity in designing electrochemical NRR systems, the solar cell area needed to produce a fixed amount of ammonia (required ammonia for a one-hectare field in the midwestern United States) is calculated.[96] If the electrochemical call can produce ammonia under 100% FE at an overpotential of 1 V, the estimated solar cell area is about 0.05% of the land area (Figure 3.4a). When the overpotential is increased to 2 V, the solar cell area is still smaller than 0.1% of the land area (Figure 3.4b). However, when the FE drops to 1%, the solar cell occupies 5% of the land (Figure 3.4c), increasing the cost of solar cells required to drive the electrochemical ammonia production.

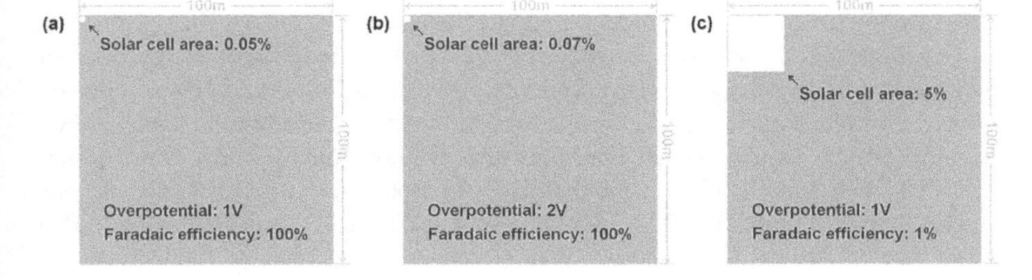

FIGURE 3.4 Minimum solar cell area needed to electrochemically produce enough ammonia for a typical one-hectare field in the midwestern United States. Figures are drawn based on the data, including overpotentials and Faradaic efficiencies published by Singh et al.[96]. Reprinted with permission from Ref.[101]

3.3 LITHIUM CYCLING STRATEGY

Lithium (Li) is the only metal in the periodic table that reacts with N_2 at room temperature to form lithium nitride (Li_3N). Then the reaction of Li_3N with proton sources (e.g., ethanol) produces ammonia. This process should be performed in aprotic solutions because Li metal reacts readily with water to release H_2 gas. Tsuneto et al.[106] proposed Li-mediated N_2 reduction, using a single compartment electrochemical cell equipped with a platinum anode and a metal cathode. The electrolyte was a THF solution containing 0.2 M $LiClO_4$ and 0.18 M ethanol as a proton source. In this process, Li^+ ions are reduced at the cathode to Li ($Li^+ \rightarrow Li$) with the cathode potential of −3 V or more negative (vs. Ag/AgCl in sat. AgCl) during electrolysis. It was demonstrated that cathode materials made of metals that electrochemically form alloys with lithium (e.g., Al, Pb) have very low NRR activities, suggesting that Li deposited on the electrode surface played a critical role in reducing the inert N_2 molecules. It is important to note that other protic compounds such as water, acetic acid, and methanol resulted in Faradaic efficiencies of less than 6.2%.[107] The main drawbacks with this system are:

i) The nitridation of the electrodeposited Li metal was conducted in the electrolyte phase. As the solubility of N_2 in most electrolytes is extremely low, high pressure is required to facilitate ammonia formation and suppress hydrogen evolution by increasing the N_2 concentration in the electrolyte.
ii) In this system, the proton donor (i.e., ethanol) was directly added to the lithium deposition solution.

Therefore, at the cathode, there is a possibility that the proton donor is reduced instead of Li^+ and the electrodeposited Li metal can directly react with the proton donor instead of undergoing nitridation. To overcome these drawbacks, the three processes of Li deposition (i.e., Li^+ reduction), Li nitridation, and ammonia formation was proposed to be physically separated from each other.[108] The proposed cycle comprised LiOH electrolysis in molten salt (LiCl and KCl) at 450 °C, followed by Li nitridation in a flowing N_2 atmosphere, and then ammonia formation by adding the Li_3N to water (i.e., hydrolysis) (Figure 3.5a). A porous alumina diffusion barrier was added around the counter electrode to mitigate the reaction of LiOH, H_2O, or O_2 with the Li product at the working electrode (Figure 3.5b). All reactions were carried out under atmospheric pressure, and near-complete conversion of Li to Li_3N was achieved after 12 h of nitridation, regardless of the reaction temperature (22, 50, and 100 °C) (Figure 3.5c). It was shown that a minimum total cell potential of ~3.0 V was required for LiOH electrolysis at 450 °C, and the FE for Li production from LiOH electrolysis was estimated to be 88.5%. This number was close to the overall FE for ammonia synthesis from N_2 and water, as Li was fully converted to Li_3N, and the Li recovery for the nitridation and hydrolysis steps was 98%[108]. The use of stainless steel cloth-based (SSC) support was proposed recently to overcome transport limitations of N_2 in tetrahydrofuran (THF) non-aqueous solution in the gas diffusion electrode setup (Figure 3.6a)[109]. The hydrophobic interactions between the electrolyte (THF solution) and support, as well as the small pore sizes in the support, prevented electrolyte

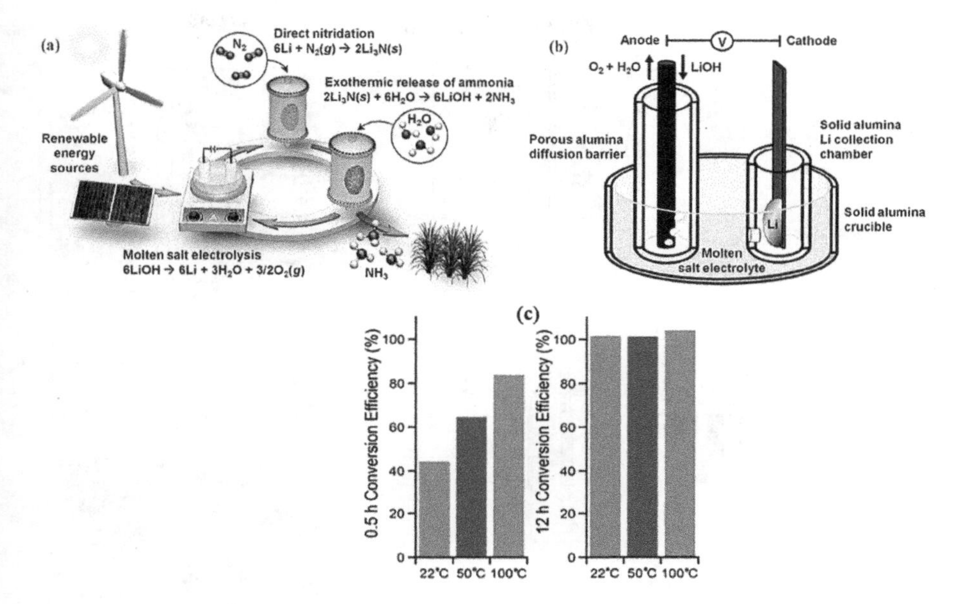

FIGURE 3.5 Li-mediated strategies for N_2 reduction to ammonia. (a) Sustainable ammonia synthesis concept cycle based on LiOH electrolysis in a molten salt electrolyte, Li nitridation, and Li_3N hydrolysis. (b) Schematic cross-section of the molten salt electrolysis cell, where a porous alumina diffusion barrier was added around the counter electrode to mitigate LiOH, H_2O, or O_2 reaction the Li product at the working electrode. (c) Ammonia yield from Li nitridation to Li_3N then reaction with H_2O at different operating temperatures (22, 50, 100 °C) for 0.5 h and 12 h experiments. Reprinted with permission from Ref.[108] Copyright 2017 the Royal Society of Chemistry.

penetration and flooding into the SSC. An ammonia partial current density of 8.8 ± 1.4 mA cm^{-2} and a FE of 35 ± 6% are obtained using a lithium-mediated approach (Figure 3.6b). In this apparatus, the NRR system was coupled with a water-splitting electrolysis cell where the produced H_2 was fed into the anode side of the electrochemical NRR system for ammonia synthesis (Figure 3.6c and 3.6d).

Electrochemical lithium cycling strategy aid in increasing the ammonia yield and FE by conducting NRR experiments in molten salt systems or non-aqueous electrolytes; however, these strategies are not energetically efficient (<5%) due to the low conductivity of non-aqueous electrolytes and/or the high temperature requirement (e.g., 450 °C) for molten salt systems.[108,109]

3.4 ELECTROCATALYTIC NRR ON AU PLASMONIC NANOPARTICLES

Theoretical and experimental studies have shown reasonable performances for electrochemical NRR using gold (Au) as an electrocatalyst.[110–112] Electrochemical NRR occurs on Au surfaces through an associative mechanism where the breaking of the triple bond of N_2 and hydrogenation of the N atoms occur simultaneously. It has been

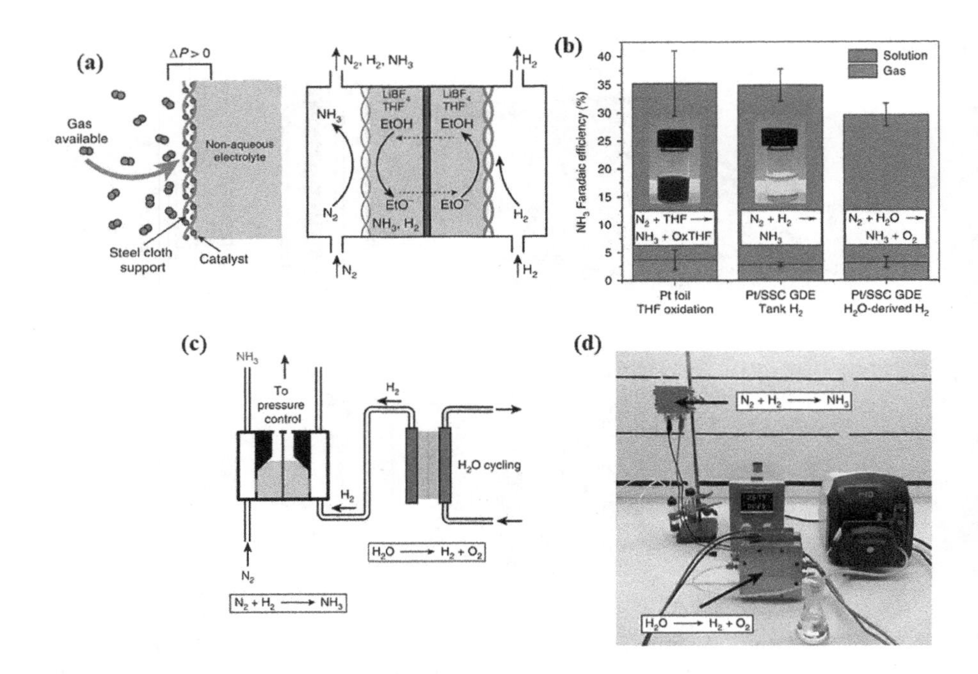

FIGURE 3.6 (a) Catalyst-coated SSC support is shown. A lack of substantial capillary action and the presence of a non-zero pressure (*P*) gradient across the cloth prevent complete catalyst flooding. Proton donor cycling is shown in a cell with a proton-producing anode. H_2 is oxidized at the anode, and nitrogen is reduced at the cathode using 1 M $LiBF_4$ in THF and 0.11 M ethanol as a proton donor. (b) Changes in FE toward ammonia with different anode chemistries in experiments where 7.2 C of charge were passed at an applied current density of 20 mA cm^{-2}, with ten standard cubic centimeters per minute (s.c.c.m.) of gas flowing past the electrode, across which the pressure gradient is one kPa. The error bars represent the standard deviation of multiple replicates of the same experiment ($n \geq 2$). The insets show the anolyte after a longer-term continuous operation at 20 mA cm^{-2} for 1 h at respective anodes. The dark solution (platinum foil) contains poorly defined THF oxidation products, whereas the clear solution (Pt/SSC) shows few signs of THF oxidation. (c) A schematic of an electrochemical HB (eHB) reactor coupled to a water-splitting reactor. (d) A photograph depicting an eHB reactor model coupled to a water electrolyzer, with the reactors indicated by arrows. Reprinted with permission from Ref.[109]

shown that N_2 adsorbs on the Au surface with further hydrogenation to form adsorbed N_2H_x species ($1 < x < 4$), where the rate-determining step is N_2 dissociation (reduction of N_2^* to form NNH^*).[110–112] Furthermore, the greater rate of NRR on gold surfaces than on the surfaces of other electrocatalysts is due to its multifaceted Au surfaces, composed of various active sites for N_2 adsorption and reduction.

Hollow Au nanoparticles can be synthesized and used in electrochemical NRR to understand the role of surface area and "confinement effect" (entrapping the reactants inside the cavity). The catalytic efficiency of AuHNCs is compared with similar concentrations of solid Au nanocubes (AuNCs), nanospheres (AuNSs), and nanorods

(AuNRs) to explore the enhanced rate of NRR using hollow nanocages. To prepare a working electrode (cathode), 300 μL of nanoparticles of known concentration and 1.5 μL of Nafion solution (5% wt.) were sonicated and dropcasted onto a substrate such as indium tin oxide (ITO) (1 cm × 1 cm) and then dried under N_2 atmosphere at 75 °C for 45 min. Electrochemical measurements were carried out at various temperatures (i.e., 20 °C, 35 °C, and 50 °C) in the water bath in 0.5 M $LiClO_4$ electrolyte (40 mL, each side) in the three-electrode setup. Pt mesh (1 cm ×1 cm) and Ag/AgCl reference electrode were used as counter and reference electrodes. A cation exchange membrane was used to separate the anodic and cathodic compartments, while protons produced at the anode can transport across the membrane to the cathode side where NRR occurs. The measured potentials vs. Ag/AgCl are iR-compensated and converted to the reversible hydrogen electrode (RHE) scale based on Equation 3.5:

$$E_{RHE} = E_{Ag/AgCl} + \frac{2.3RT}{F} pH + E^{o}_{Ag/AgCl} \tag{3.5}$$

where E_{RHE} is the converted potential vs. RHE, $E^{o}_{Ag/AgCl} = 0.2027$ at 20 °C with the slope of −1.01 mV/°C, $E_{Ag/AgCl}$ is the experimentally measured potential against Ag/AgCl reference electrode, R is the gas constant (8.314 J mol^{-1} K^{-1}), and T is the operating temperature (K). To calculate the ammonia FE, the amount of ammonia produced during the experiment is divided by the total charge applied to the electrodes. Based on this reaction ($N_2 + 6H^+ + 6e^- \rightarrow 2NH_3$), three electrons are required to produce one mole of NH_3. The FE is calculated according to Equation 3.6:

$$FE_{NH3}(\%) = \frac{C \times V}{(i \times t)/(n \times F)} \tag{3.6}$$

where C is the ammonia concentration (mol L^{-1}), V is the volume of the electrolyte (L), i is the measured current (A), n is the number of electrons that are required to produce one mole of ammonia (eq. mol^{-1}), and F is the Faraday's constant (96,485 C eq^{-1} or A.S eq^{-1}). The molar concentration of each nanoparticle (C) is determined by dividing the total number of gold atoms (N_{total}) in the sample over the average number of gold atoms per nanoparticle (N) according to Equation 3.7:

$$C = \frac{N_{total}}{NVN_A} \tag{3.7}$$

where V is the volume of the sample, and N_A is the Avogadro's constant (6.022×10^{23} mol^{-1}).

The concentration of nanoparticles is measured using inductively coupled plasma emission spectroscopy (ICP-ES). The weight of the nanoparticle is determined from the high-resolution TEM image. To calculate N, the mass of each nanoparticle ($\rho V'$) is divided to the mass of one gold atom (M) based on Equation 3.8:

$$N = \frac{\rho V'}{M} \tag{3.8}$$

where ρ is the density of Au atom (19.3×10^{-21} g/nm^3), V' is the volume of each nanoparticle, and M is the mass of one gold atom, which is the ratio of the molecular weight of Au (197 g mol^{-1}) and Avogadro's constant.

$$Mass\ of\ one\ gold\ atom\left(M\right)=\frac{197\,g\,mol^{-1}}{6.022\times10^{-23}\,atom\,mol^{-1}}=3.27\times10^{-22}\frac{g}{Au\,atom} \tag{3.9}$$

To calculate the mass of each nanoparticle, the diameter of a nanoparticle is determined from the TEM image. For instance, the average diameter of Au nanospheres (AuNSs) is 35 nm. Therefore, the mass of the nanosphere is calculated by multiplying the volume of the nanosphere ($\pi d^3/6$) and the density of Au according to Equation (3.10):

$$Mass\ of\ AuNS=19.3\times10^{-21}\frac{g}{nm^3}\times\pi\times\frac{35^3}{6}=4.33\times10^{-16}\,g \tag{3.10}$$

Then, the number of Au atoms per nanoparticle is calculated by dividing Equations (3.9) and (3.10):

$$N=\frac{Mass\ of\ AuNS}{Mass\ of\ one\ gold\ atom}=\frac{4.33\times10^{-16}\,g}{3.27\times10^{-22}\frac{g}{Au\,atom}}=1{,}324{,}989.119\,Au\,atom \tag{3.11}$$

Finally, the value of Au concentration obtained from ICP-ES (e.g., for AuNSs: 8.28 mg L^{-1}) is divided by the average number of Au atoms per nanoparticle to determine the concentration of nanoparticles in the sample, which is ~32 pM.

AuHNCs are prepared from a solid silver nanocubes (AgNCs) template by the galvanic replacement technique.[5,113] In this method, the sacrificial metal template (i.e., Ag) is replaced with the nanocage metal (i.e., Au) if the oxidation potential of the metal template is higher than that of the nanocage metal. The replacement of three Ag atoms of the template with one Au atom ($3Ag(s) + AuCl_4^-(aq.) \rightarrow Au(s) + 3Ag^+(aq.) + 4Cl^-(aq.)$) creates a hollow structure with holes at the wall and corners of the nanocage. The size of the AuHNCs is tuned by varying the size of the AgNCs. AgNCs with localized surface plasmon resonance (LSPR) peak position at 429 nm are prepared by a modified polyol reduction method (Figure 3.7a).[114] The AgNCs are then washed with acetone, centrifuged (for 10 min at 10,000 rpm) and dispersed in DI water. Hydrogen tetrachloroaurate (0.5 mM) in DI water is injected into the AgNC solution under vigorous stirring (600 rpm) until the LSPR peak position shifts to 660 nm (Figure 3.7a). The X-ray diffraction (XRD) pattern of AuHNCs deposited on the Si substrate is composed of various surface index facets that provide active sites for electrochemical NRR (Figure 3.7b). The average edge length of AuHNCs is 35 nm, obtained from transmission electron microscopy (TEM) analysis (Figure 3.7c

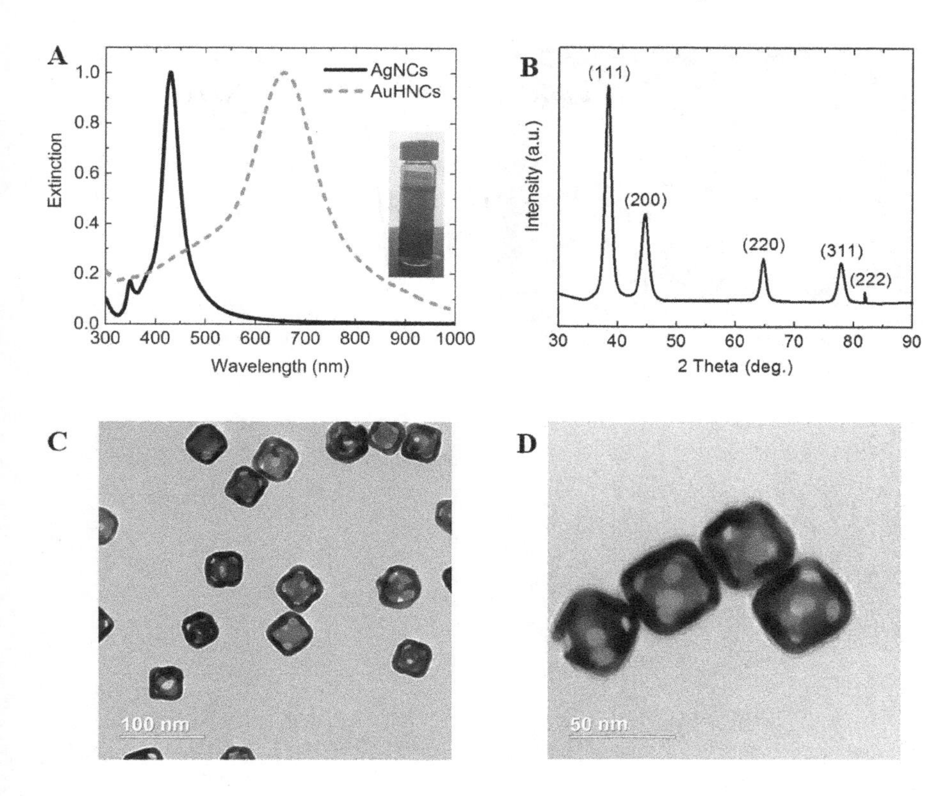

FIGURE 3.7 (a) UV-vis extinction spectra of the AgNCs and AuHNCs, the photograph shown in the inset is AuHNCs dispersed in DI water. (b) XRD pattern of AuHNCs deposited on Si substrate. (c and d) TEM images of AuHNCs with different magnifications. The average edge length of the AuHNCs is 35 nm. Reprinted with permission from Ref.[114]

and 3.7d). Electrochemical NRR experiments are conducted in H-type cells where anodic and cathodic compartments are separated by a proton conductive cation exchange membrane (Figure 3.8a). The electrolyte is 0.5 M $LiClO_4$ aqueous solution. Perchlorate anion (ClO_4^-) is beneficial for NRR due to the minimal and unselective adsorption of the ClO_4^- on the low index facets of Au nanoparticles surface[115]. Li^+ is advantageous for its superior ability to activate N_2 at ambient conditions.[85,108,116]

Linear sweep voltammetry (LSV) in an Ar and N_2 saturated environment qualitatively distinguished between HER and NRR (Figure 3.8b). As the potential moves below −0.4 V vs. RHE, a notable enhancement in current density is observed under the N_2 saturated environment. This is attributed to the reaction between AuHNCs as a cathodic electrocatalyst and N_2 to produce NH_3. The greatest difference in the LSV curves of N_2 and Ar-saturated environments is found at the potential between −0.4 V to −0.8 V vs. RHE. It is expected that the highest NRR activity in this potential range was achieved. Moving toward potentials more negative than −0.8 V vs. RHE yields no further difference in current density between Ar and N_2 (Figure 3.8b). This indicates that HER is the only reaction at the cathode. Chronoamperometry (CA) tests at

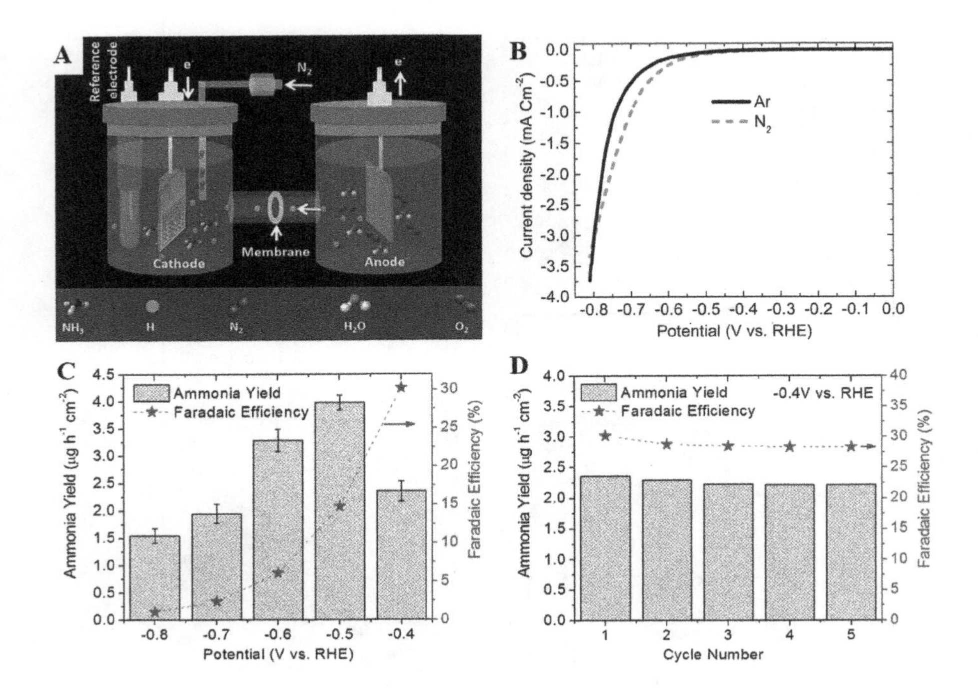

FIGURE 3.8 (a) Schematic of electrochemical cell for NRR. The anode and cathode compartments are separated by a cation exchange membrane (CEM). (b) LSV tests in an Ar and N_2 saturated environment in 0.5 M $LiClO_4$ aqueous solution under ambient conditions. (c) Ammonia yield rate and FE at various potentials in 0.5 M $LiClO_4$ at 20 °C. (d) Cycling stability results of ammonia yield rate on AuHNCs. For each cycle CA test was carried out at −0.4 V vs. RHE in 0.5 M $LiClO_4$ at 20 °C. Reprinted with permission from Ref.[114]

a series of potentials were conducted to determine the ammonia yield rate and FE. The calibration curve for the ammonia assay using the Nessler reagent is shown in Figure 3.9. The electrochemical NRR obtains higher selectivity within the potential range of −0.4 V to −0.6 V vs. RHE, with the highest ammonia yield rate (3.98 μg cm^{-2} h^{-1}) at −0.5 V and FE (30.2%) at −0.4 V. Although in this potential range (−0.4 V to −0.6 V), the ammonia yield rate increases as the negative applied potential increases, FE decreases, which is attributed to the compromise between increasing current density and competitive selectivity toward HER rather than NRR. It is known that H atoms occupy the active sites on Au that prevents N_2 adsorption and reduction on the catalytic surface.[111,117] At negative applied potential below −0.6 V, both ammonia yield rate and FE decrease considerably (ammonia yield rate: 1.54 μg cm^{-2} h^{-1}, FE:1.1% at −0.8 V vs. RHE), which suggests that HER is the dominant reaction at the cathode (Figure 3.8c). To evaluate the durability of AuHNCs, CA tests are performed at −0.4 V vs. RHE for 5 consecutive cycles, each for 12 h. The electrocatalyst shows excellent stability with a minor decrease in ammonia yield rate and FE (93.8% performance retention) after the 5th recycling test (Figure 3.8d).[114]

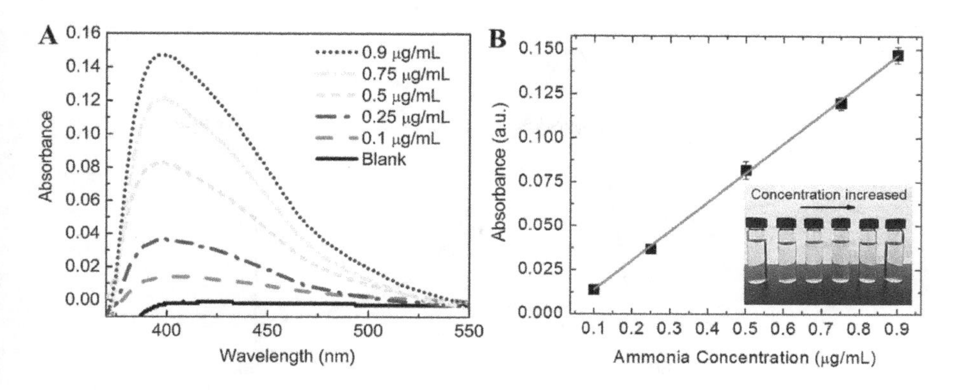

FIGURE 3.9 (a) UV-vis calibration curve for ammonia quantification using Nessler's method. Known concentration of ammonium ions are added to 0.5 M $LiClO_4$ electrolyte and mixed thoroughly with 1 mL of 0.2 M $KNaC_4H_6O_6$ and 1 mL of Nessler reagent and then the absorbance at 395 nm is measured by the UV-vis spectrophotometer. The value of blank electrolyte is subtracted from all other concentrations as background. Reproduced with permission from Ref.[114]

To further verify the electrochemical NRR activity using AuHNCs, control experiments are carried out with Ar gas and with no potential applied to the electrodes under N_2 gas (open circuit voltage). Significantly smaller amounts of NH_3 yield are measured using Ar gas or N_2 gas without applying a potential (Figure 3.10a). This confirms that the results with the N_2 gas under applied potential are not due to the sources of contamination (e.g., laboratory, equipment, membrane). During the electrochemical NRR experiments, the formation of ammonia is further validated by surface-enhanced Raman spectroscopy (SERS).[118] The solution collected from the electrochemical experiment, which showed the highest ammonia FE (−0.4 V vs. RHE at 20 °C), was mixed with AuNCs, and the SERS spectrum was collected under 785 nm laser excitation (Figure 3.10b). A control experiment was performed, where the SERS spectrum for 0.5 M $LiClO_4$ was collected under the same operating condition, as the SERS band might correspond to the ligand molecules on the AuNCs surface and $LiClO_4$ could interfere with the Raman features of ammonia. In comparison to the SERS spectra collected from the control experiment, a new Raman band at 3062 cm^{-1} appeared for the sample analyzed after the electrochemical experiment, which is attributed to the NH_4^+ moiety present in the electrolyte.[119,120] This was further validated by collecting the Raman spectrum from crystalline NH_4F, which showed a corresponding vibration mode for NH_4^+ at around 3075 cm^{-1}. The observed shift in the Raman band corresponds to NH_4^+ in the SERS spectrum compared to the normal Raman spectrum could be attributed to the plasmon-enhanced electromagnetic field effect and the modifications in the adsorption (physisorption and chemisorption) orientation of these molecules at the nanoparticle surface. This result points toward the fact that ammonia formed during the electrochemical experiment primarily exists as ammonium ions (NH_4^+) in the electrolyte. Furthermore, ^1H NMR spectra obtained from the sample in the NRR experiment lie at a chemical shift of triplet

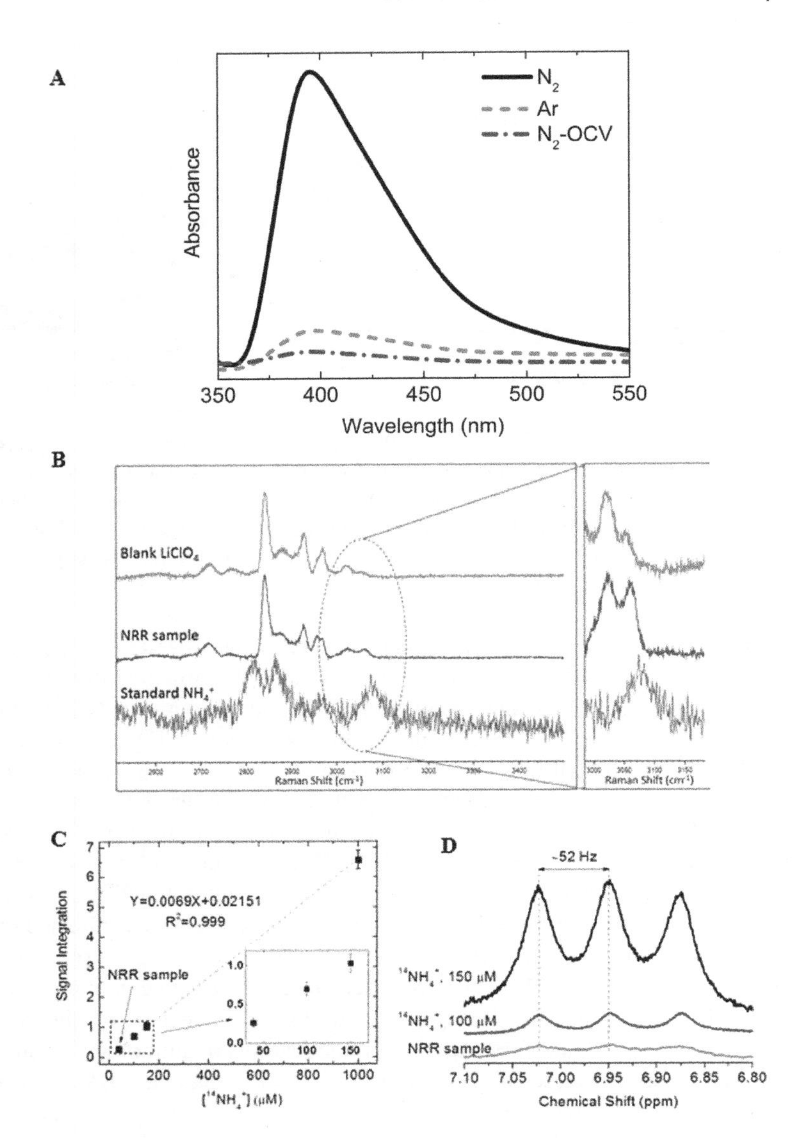

FIGURE 3.10 (a) UV-vis absorption spectrum of AuHNCs in 0.5 M $LiClO_4$ aqueous solution under N_2 saturated with and without applied potential and Ar gas with applied potential (−0.4 V vs. RHE). (b) Raman spectra of blank 0.5 M $LiClO_4$ aqueous solution, NRR sample, and standard NH_4^+ sample. (c) Calibration curve of the ^{1}H NMR signal at 6.95 ppm for standard solutions of NH4+ (100, 150, 1000 μM) and NRR sample. (d) ^{1}H NMR spectra (700 MHz) of NH_4^+ produced from the NRR experiment and standard NH_4^+ samples. ^{14}N produces a triplet with ~52 Hz J coupling constant. Reprinted with permission from Ref.[114]

coupling of $^{14}N_2$ similar to that of standard $^{14}NH_4^+$ samples (J- coupling: ~52 Hz), which further confirms that the ammonia formation is solely originated from N_2 (Figure 3.10c). The ammonia production at −0.4 V vs. RHE at 20 °C quantified from the ^{1}H NMR analysis is 38.6 μM (Figure 3.10d). This is compared with 41.5 μM of ammonia, measured by UV-vis spectra using Nessler's test.

By increasing the concentration of AuHNCs from 0.9 μg mL^{-1} to 1.8 μg mL^{-1} on the Si substrate (1 cm^2) at −0.5 V vs. RHE, both the ammonia yield rate and the FE increase from 1.88 μg cm^{-2} h^{-1} and 3.12% to 3.98 μg cm^{-2} h^{-1} and 14.8% (Figure 3.11a). This is due to the increase in the number of nanoparticles participating in the NRR, resulting in an increase in the total active surface area. By increasing the electrochemical NRR temperature from 20 °C to 50 °C at −0.4 V vs. RHE, the ammonia yield rate and FE increase from 2.35 μg cm^{-2} h^{-1} and 30.22% to 2.82 μg cm^{-2} h^{-1} and 40.55% (Figure 3.11b). Based on the Arrhenius equation, the reaction rate increases exponentially with the temperature. The mass transport rate is faster at higher temperatures, while the N_2 solubility decreases at higher temperatures. For these experiments, faster kinetics plays a significant role in enhancing the NRR rate at higher temperatures. The ammonia yield rate and FE can be further enhanced by using ionic liquids (e.g., [C4mpyr][eFAP]) that offer significantly higher N_2 solubility compared to the aqueous solution.[121]

To further investigate the cage effect using AuHNCs, the NRR rate is evaluated using solid Au nanoparticles of various shapes (i.e., rods, spheres, cubes) with similar nanoparticle concentrations. The LSPR of AuNSs and AuNCs lies at 535 nm and two plasmon peaks for AuNRs are observed, which are attributed to the transverse (512 nm) and longitudinal (746 nm) modes (Figure 3.12a). The average diameter and edge length of AuNSs and AuNCs, respectively, are 35 nm, while AuNRs have an average length and width of 42 nm and 12 nm as obtained from TEM images (Figure 3.12b–d). The ammonia yield rate and FE are significantly lower using solid Au nanoparticles compared to the AuHNCs (Figure 3.12e). The lowest ammonia yield rate and FE are obtained for AuNRs (0.99 μg cm^{-2} h^{-1}, 10.69%). A minor increase in the NRR rate is observed using AuNSs (ammonia yield rate: 1.19 μg cm^{-2} h^{-1}, FE: 11%) and AuNCs (ammonia yield rate: 1.27 μg cm^{-2} h^{-1},

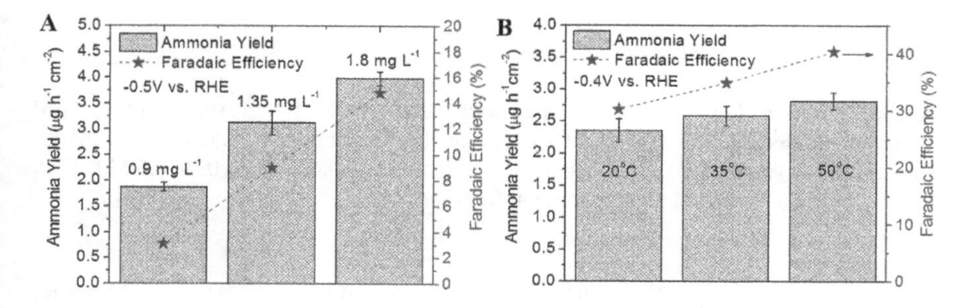

FIGURE 3.11 (a) Ammonia yield rate and FE of AuHNCs at various concentrations at −0.5 V vs. RHE in 0.5 M $LiClO_4$ aqueous solution under ambient conditions. (b) Ammonia yield rate and FE of AuHNCs at various temperatures at −0.4 V vs. RHE in 0.5 M $LiClO_4$ aqueous solution. Reprinted with permission from Ref.[114]

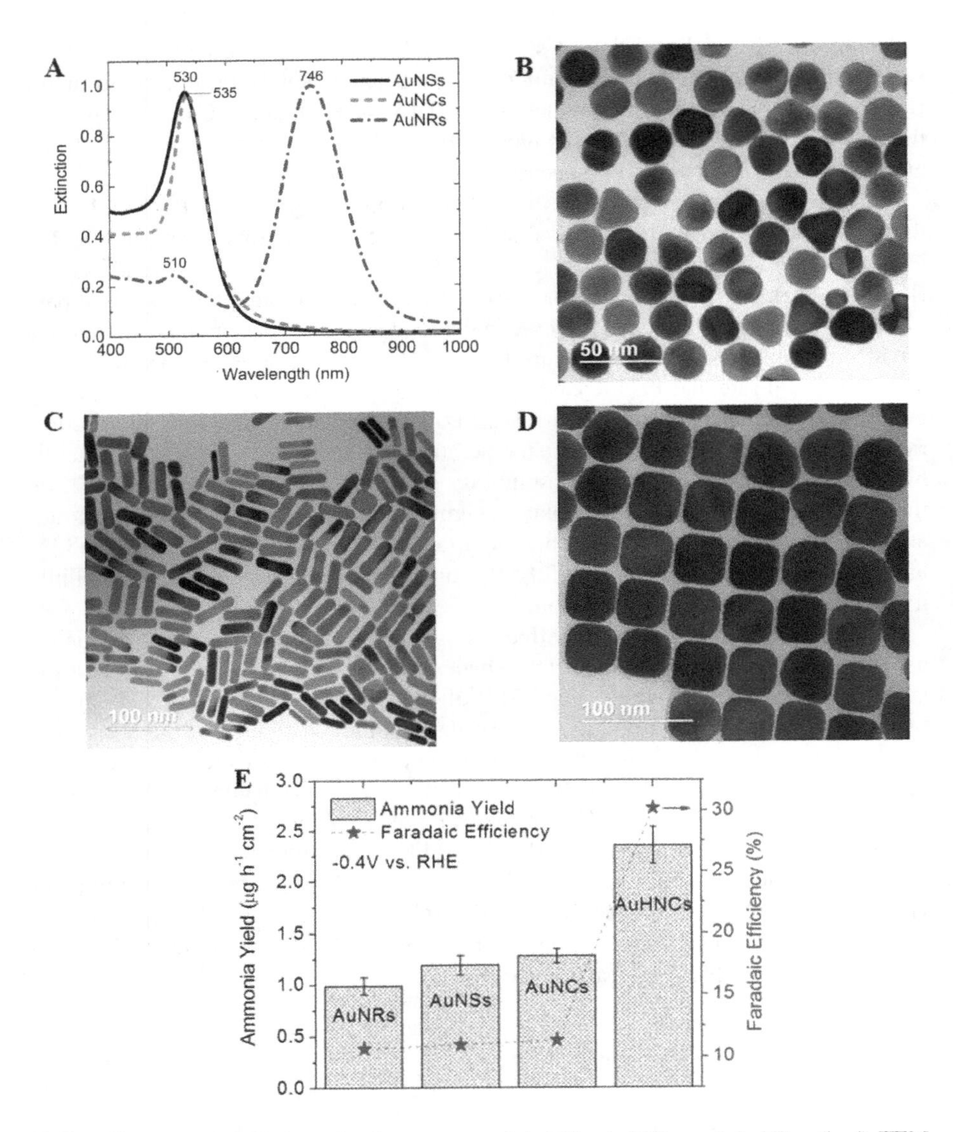

FIGURE 3.12 (a) UV-vis extinction spectra of AuNSs, AuNCs, and AuNRs. (b–d) TEM images of AuNSs, AuNRs, and AuNCs, respectively. (e) Ammonia yield rate and FE for nanoparticles of various types and shapes at the potential of −0.4 V vs. RHE in 0.5 M $LiClO_4$ aqueous solution. Reprinted with permission from Ref.[114]

FE: 11.35% for AuNCs). Nanoparticles with sharper edges and corners yield higher catalytic activity.[13] This is due to the increased number of valency-unsatisfied surface atoms (atoms that do not have the full number of bonds that they can chemically accommodate) in nanoparticles with sharper edges, providing more active sites for catalytic reaction than smoother nanoparticles.[13] The significant enhancement in the

NRR rate using AuHNCs is attributed to the entrapment of N_2 molecules within the cavity. These likely experience high-frequency collisions with the hollow Au interior surface of the cages. This increases the residence time of N_2 molecules at the inner nanoparticle surface, which facilitates the conversion of N_2 to NH_3. In addition, the presence of less capping material (i.e., polyvinylpyrrolidone (PVP)) on the inner surface than the outer surface of the AuHNCs can enhance the NRR catalytic activity.

3.5 ELECTROCATALYTIC NRR USING PORE-SIZE CONTROLLED HOLLOW BIMETALLIC AU–AG NANOCAGES

Finding the optimum size and density of pores in the walls of AuHNCs is crucial to enhance the rate of electro-reduction of N_2 to NH_3. The interdependency between the pore size/density, the LSPR peak position, the silver content in the cavity, and the nanoparticle's total surface area should be explored for further optimization of hollow plasmonic nanocatalysts in electrochemical NRR. The electrochemical surface areas (ECSA) of AuHNCs with various pore sizes are determined by cyclic voltammetry (CV) tests in Ar-saturated 0.1 M LiOH aqueous solution in a rotating disk electrode (RDE) setup at a scan rate of 50 mV s^{-1}.[122] A polished glassy carbon disk electrode mounted on an interchangeable RDE holder was used as the working electrode. A platinum coil and a single junction Ag/AgCl reference electrode (4 M KCl with AgCl solution) were used as counter and reference electrodes. To mitigate any interferences on CV measurements, both counter and reference electrodes were separated from the main cell by an electrolyte bridge. Before measuring CVs for $ECSA_{Au}$ calculation, the working electrode is conditioned by conducting CV tests for 50 cycles at a scan rate of 200 mV s^{-1} to remove possible surface impurities and achieving a stable current density response.[123] Then, 20 μL of AuHNCs with 0.1 μL of Nafion solution (5% wt.) are sonicated and dispersed on a polished glassy carbon electrode with Au loadings of 0.12, 0.19, and 0.27 μg_{Au} $cm^{-2}{}_{disk}$ for AuHNCs-635, AuHNCs-715, and AuHNCs-795. To more uniformly disperse nanoparticles on the glassy carbon, nanoparticles are dried on the glassy carbon with a rotation rate of 600 rpm at room temperature. The $ECSA_{Au}$ of the catalyst is determined from the charge associated with the reduction peak of Au oxide after double-layer correction and is normalized to the Au loading on the working electrode and charge density of 386 μC cm^{-2} according to Equation 3.12:

$$ECSA\left(\frac{cm_{Au}^{2}}{g_{Au}}\right) = \frac{Q\left(\mu C\, cm^{-2}\right)}{386\, \mu C\, cm_{Au}^{-2} \times electrode\ loading\left(g_{Au} cm^{-2}\right)} \tag{3.12}$$

where Q (μC cm^{-2}) is the charge associated with the reduction peak of Au oxide after double-layer correction and calculated according to the Equation (3.13):

$$Q = \frac{\int iV}{\upsilon} \tag{3.13}$$

where i is the current density (μA cm^{-2}), V is the potential (V), and v (V s^{-1}) is the scan rate.

AuHNCs with various LSPR peak values (i.e., 635 nm, 715 nm, and 795 nm) are prepared by adding hydrogen tetrachloroaurate (0.5 mM $HAuCl_4$ (aq.)) in solid Ag nanocubes (AgNCs) that are dispersed in DI water using the synthetic galvanic method.[21] By increasing the amount of Au^{3+} ions added to the AgNCs template with the initial LSPR peak position at 445 nm, the core Ag atoms are etched, and the resulting LSPR peak value of AuHNCs redshifts (Figure 3.13a). The first stage of this synthesis after adding Au^{3+} to the template solution is the formation of nanoboxes with walls composed of Ag–Au alloy at the LSPR peak value of 635 nm (Figure 3.13b). As more Au^{3+} is added to the boxlike AuHNCs solution, the de-alloying process of Ag atoms from the Ag–Au walls is initiated. Numerous pores are formed at the walls and corners of AuHNCs at the LSPR peak value of 715 nm (Figure 3.13c). By further adding Au^{3+} to the porous AuHNCs, the pore size increases while the pore density decreases and the LSPR peak value redshifts to 795 nm (Figure 3.13d). By shifting the LSPR peak value from 635 nm to 715 nm, the small

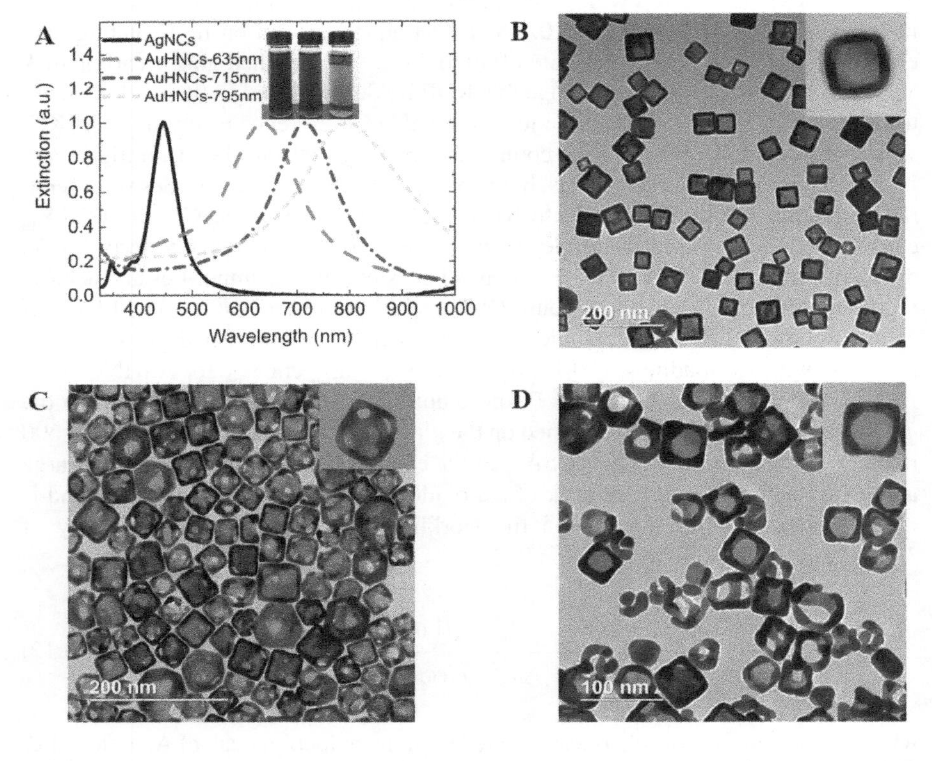

FIGURE 3.13 (a) UV-vis extinction spectra of AgNCs and AuHNCs with various LSPR peak values, the photograph shown in the inset is AuHNCs dispersed in DI water. (**b–d**) TEM images of AuHNCs with LSPR peak values at 635 nm, 715 nm, and 795 nm, respectively. The inset of each image is the magnified TEM image of a nanoparticle. Reprinted with permission from Ref.[123]

peak in UV-vis spectrum around 445 nm disappears, indicating the removal of Ag from the interior surface of the cavity (Figure 3.13a). The Au and Ag concentrations of all synthesized nanoparticles are determined by inductively coupled plasma emission spectroscopy (ICPES) and reported in Table 3.1. It is observed Au content (mass %) in nanoparticles increases from 33.0 to 64.7 as the LSPR peak position shifts from 635 nm to 795 nm (Table 3.1). The $ECSA_{Au}$ is calculated from the reduction peak of Au oxide after double-layer correction and a charge density of 386 μC cm^{-2} (Figure 3.14 and Table 3.1). AuHNCs-715 has the highest $ECSA_{Au}$ (26.6 m^2 g^{-1}) while AuHNCs-795 has the lowest $ECSA_{Au}$ (21.3 m^2 g^{-1}). Although AuHNCs-795 has the highest Au concentration among all nanoparticles (Table 3.1), it has the lowest $ECSA_{Au}$. This is attributed to the simultaneous reduction of Au and Ag atoms in the final stage of synthesis when the LSPR redshifts from 715 nm to 795 nm, which changes the nanocages' porosity and increases the void size. The smaller peak observed for the reduction of Au oxide compared to similar studies[122,124] is due to the significantly smaller Au loading (e.g., 0.39 μg_{Au} cm^{-2} disk for AuHNCs-715), which is necessary when the economic feasibility of using this electrocatalyst for ammonia synthesis is investigated.[123]

LSV tests are carried out in Ar and N_2 saturated electrolyte to evaluate the selectivity performance $(I_{N2} - I_{Ar})/(I_{N2}) \times 100$ of AuHNCs with various values of LSPR peak toward NRR. Faradaic current is obtained by subtracting the capacitive current from the actual current recorded from LSV tests. For all electrocatalysts, current density differs between Ar and N_2 within the potential window of −0.3 V to −0.6 V

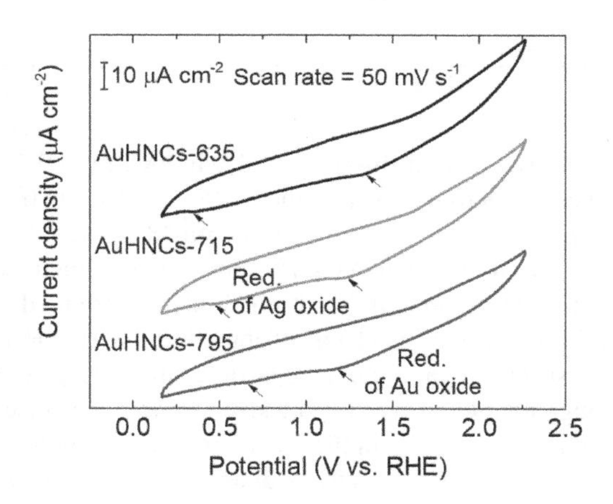

FIGURE 3.14 Cyclic voltammograms (CV) of AuHNCs with various peak LSPR values in Ar-saturated 0.1 M LiOH aqueous solution at a scan rate of 50 mV s^{-1}. The CV measurements were conducted in the RDE setup at a rotation rate of 1500 rpm at room temperature. The observed shift in decreasing the reduction potential of Au oxide (~0.15 V) when LSPR redshifts from 635 nm to 795 nm is attributed to the de-alloying process by removal of Ag in the cavity. The second peak in the reduction segment of the CV curve corresponds to the reduction of Ag oxide. The intensity of the peak is proportional to the Ag concentration which is the highest for AuHNCs-715. Reprinted with permission from Ref.[123]

TABLE 3.1
Au, Ag Concentrations, Au and Ag Content (% at.) of Nanoparticles are Determined by ICPES. The Electrochemical Surface Areas (ECSA) of Nanoparticles are Determined based on the Reduction Peak of Au Oxide during CV Measurement in Ar-Saturated 0.1 M LiOH Solution at a Scan Rate of 50 mV s^{-1}. Atomic Content is Calculated using Au and Ag Concentrations Divided by the Molar Mass of Au (196.97 g mol^{-1}) and Ag (107.87 g mol^{-1})

Catalyst	Au Conc. (μg mL^{-1})	Ag Conc. (μg mL^{-1})	Au Content (Mass %)	Au Content (Atom %)	$ECSA_{Au}$ (m^2 g^{-1})
AuHNCs-635	1.20	2.44	33.0	21.2	23.4
AuHNCs-715	3.91	3.45	53.1	38.3	26.6
AuHNCs-795	5.40	2.95	64.7	50.1	21.3

TABLE 3.2
Selectivity Performance of AuHNCs with Various LSPR Peak Values Toward NRR. Reprinted with Permission from Ref[123]

	Potential (V vs. RHE)		
Electrocatalyst	−0.3	−0.4	−0.5
AuHNCs-635 nm	52.3	48.7	45.1
AuHNCs-715 nm	55.6	65.3	56.3
AuHNCs-795 nm	20.2	27.4	11.0

vs. RHE (Figure 3.15a). Beyond this potential window, the HER is the dominant reaction. The highest selectivity toward the NRR (65.3% at−0.4 V vs. RHE) is achieved using AuHNCs-715. This is compared with 48.7% and 27.4% at −0.4 V vs. RHE for AuHNCs-635 nm and AuHNCs-795 nm. The complete selectivity performance of electrocatalysts at three different potentials is provided in Table 3.2. The highest N_2 selectivity for AuHNCs-715 is attributed to the compromise between the pore size, the active surface area of the nanoparticle, and the Ag content in the cavity of AuHNCs. Although by increasing the pore size, the active surface area of nanoparticles decreases, the presence of Ag in the cavity of AuHNCs with smaller pore sizes (i.e., AuHNCs-635) decreases the selectivity of an electrocatalyst toward NRR, as Ag enhances H_2 evolution. It is noted that all electrochemical experiments are conducted in an O_2-free environment to avoid the formation of silver (I) oxide Ag_2O in the cavity. As the NRR primarily happens within the cavity, the AuHNCs that have smaller pore sizes but that have higher Ag within the cavity (AuHNCs-635) result in lower selectivity for NRR compared with AuHNCs that have bigger pore size but have lower Ag in the cavity (AuHNCs-715). Further increasing the pore size decreases the NRR selectivity due to both the surface area's decrease and the inefficient confinement of reactants within the cavity (AuHNCs-795). The pore size should be

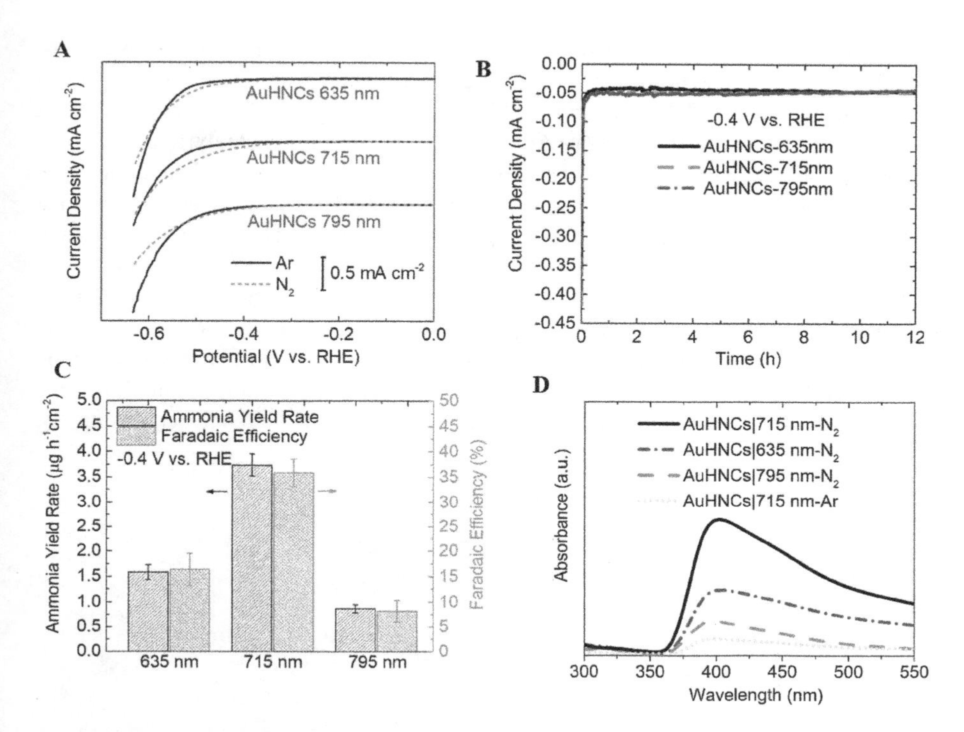

FIGURE 3.15 (a) LSV tests of AuHNCs with LSPR peak values at 635 nm, 715 nm, and 795 nm in an Ar and N_2 saturated 0.5 M $LiClO_4$ (aq.) under ambient conditions with the scan rate of 10 mV s^{-1}. (b) Chronoamperometry (CA) results of AuHNCs with various LSPR peaks at −0.4 V vs. RHE in N_2 saturated 0.5 M $LiClO_4$ (aq.). (c) Ammonia yield rate and FE for AuHNCs with various LSPR peak values at the potential of −0.4 V vs. RHE in 0.5 M $LiClO_4$ (aq.). (d) UV-vis absorption spectra of N_2 and Ar-saturated 0.5 M $LiClO_4$ (aq.) after electrolysis at −0.4 V vs. RHE for 12 h using Nessler's test for AuHNCs with various LSPR peak values. Reprinted with permission from Ref.[123]

engineered so that reactants can diffuse in and products can diffuse out of the cavity while not losing the surface area notably, due to the increase in the pore size in the walls of the nanocages. Therefore, the pore size optimization plays a crucial role in confining the reactants in the small region within the cavity and increasing their collision frequency with the interior surface of the electrocatalyst.

Chronoamperomtry (CA) tests are conducted at −0.4 V vs. RHE to evaluate the electrocatalytic activity and determine the ammonia yield rate and FE for AuHNCs-635, AuHNCs-715, and AuHNCs-795 (Figure 3.15b). The highest ammonia yield rate (3.74 μg cm^{-2} h^{-1}) and FE (35.9%) are achieved using AuHNCs-715, while the lowest ammonia yield rate (0.87 μg cm^{-2} h^{-1}) and FE (8.2%) are obtained for AuHNCs-795 (Figure 3.15c). This is in line with the trend of electrocatalysts' selectivity performance for NRR (Figure 3.15a). A small amount of NH_3 is detected (~0.34 μg cm^{-2} h^{-1}, ~9% of N_2 gas) when N_2 gas is replaced with Ar under the same applied potential using AuHNCs-715, which indicates that the majority of NH_3 in these experiments originated from the N_2 source (Figure 3.15d). Hollow Au nanospheres (AuHNSs) with LSPR peak values and electrolysis operating conditions similar to those of AuHNCs (Figure 3.15c) are evaluated for electrochemical NRR (Figure 3.16). The same trend as AuHNCs but lower electrocatalytic activity for NRR is observed using AuHNSs. The highest ammonia yield rate (2.77 $\mu g\ cm^{-2}\ h^{-1}$) and FE (29.3%) are achieved using AuHNSs-715; these values are lower than AuHNCs-715 (3.74 $\mu g\ cm^{-2}\ h^{-1}$, FE = 35.9%) (Figure 3.17). This could be attributed to small variations in the nanospheres' sizes, cavity volume, pore sizes, and lack of sharp corners and edges in AuHNSs. The role of pH on the electrocatalytic activity of NRR is evaluated using AuHNCs-715. It is found that operating the N_2 electrolysis in both acidic (0.5 M $LiClO_4$ + 0.001 M $HClO_4$, pH = 3) and alkaline electrolyte (0.1 M LiOH, pH = 13) decreases the rate of ammonia production (Table 3.3). This is attributed to the favorable HER in acidic conditions. Moreover, decreasing the

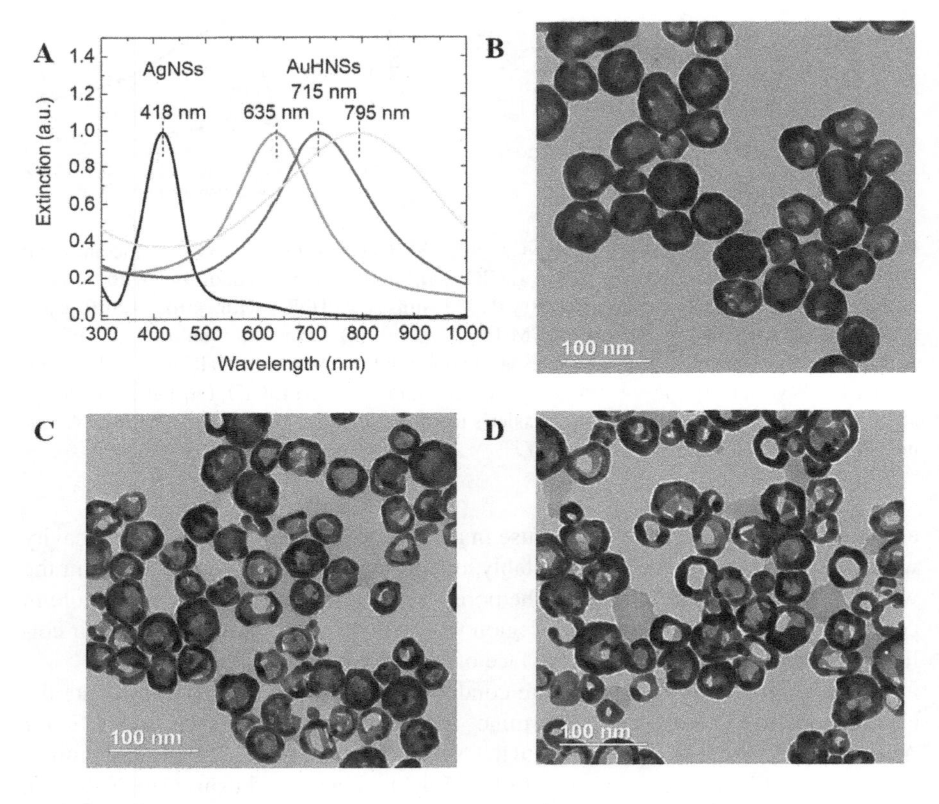

FIGURE 3.16 (a) UV-vis extinction spectra of AgNSs and AuHNSs with various LSPR peak values. (b–d) TEM images of AuHNSs with the LSPR peak values at 635 nm, 715 nm, and 795 nm, respectively. The average diameter of nanoparticles is 55 nm. Reprinted with permission from Ref.[123]

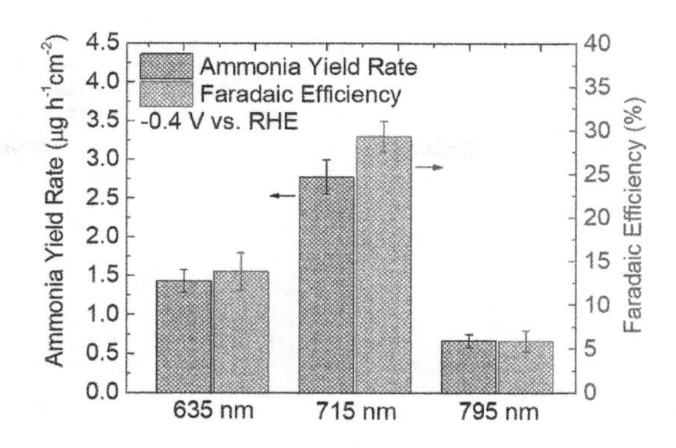

FIGURE 3.17 Ammonia yield rate and FE for AuHNSs with various LSPR peak values at the potential of −0.4 V vs. RHE in 0.5 M $LiClO_4$ (aq.). Reprinted with permission from Ref.[123]

TABLE 3.3
Electrochemical Performance of NRR using AuHNCs-715 with Various pH Electrolytes at −0.4 V vs. RHE

Electrolyte	NH_3 Yield Rate ($\mu g\ cm^{-2}\ h^{-1}$)	Average Current Density at −0.4 V vs. RHE ($\mu A\ cm^{-2}$)	FE (%)
0.5 M $LiClO_4$ + 0.001 M $HClO_4$ (pH = 3)	3.10 ± 0.15	57.1	25.7
0.5 M $LiClO_4$ (pH = 8)	3.74 ± 0.22	49.2	35.9
0.1 M LiOH (pH = 13)	0.71 ± 0.06	46.2	7.2

Reprinted with permission from Ref.[123]

concentration of Li^+ in the alkaline electrolyte (0.1 M) compared with neutral and acidic electrolytes (0.5 M) results in a remarkable decrease of the NH_3 yield rate (Table 3.3). It is noted that an anion exchange membrane (AEM) is used for the electrolysis of N_2 in the alkaline solution to enable the transport of hydroxide anions (OH^-) from the cathode to the anode side for stable electrolysis.

CA tests are performed using AuHNCs-715 to determine the ammonia yield rate and FE at a series of applied potentials (Figures 3.18a and 3.18b). The highest ammonia yield rate (4.22 $\mu g\ cm^{-2}\ h^{-1}$) is obtained at −0.5 V vs. the RHE with the FE of 17.9%. The higher ammonia yield rate but lower FE at −0.5 V, compared with −0.4 V, is due to the compromise between achieving higher current density and higher selectivity toward HER. Moving toward more positive potentials than −0.4 V and more negative potentials than −0.5 V, both NH_3 yield rate and FE decrease,

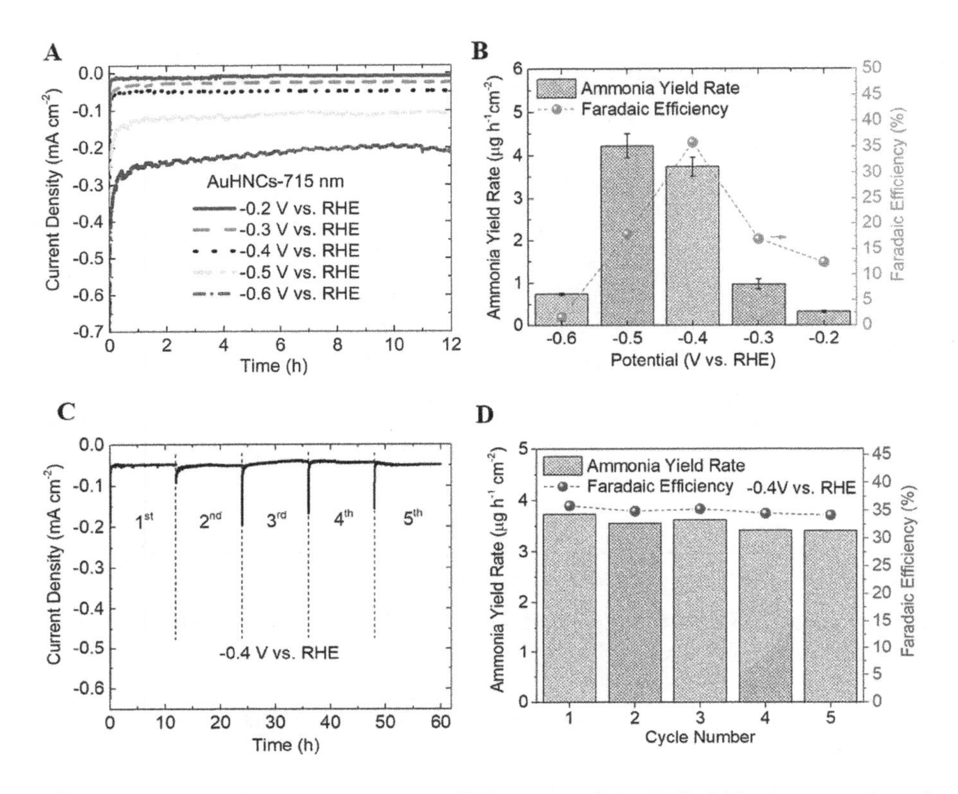

FIGURE 3.18 (a) Chronoamperometry (CA) results of AuHNCs-715 nm at a series of potentials. (b) Ammonia yield rate and FE at various potentials in 0.5 M $LiClO_4$ (aq.). (c) CA tests for the stability of the AuHNCs-715 nm at −0.4 V vs. RHE in 0.5 M $LiClO_4$ (aq.). For each cycle, a CA test was carried out at −0.4 V vs. RHE for 12 h. (d) Cycling stability results of ammonia yield rate and FE on AuHNCs-715 nm. For each cycle, a CA test was carried out at −0.4 V vs. RHE in 0.5 M $LiClO_4$ (aq.). Reprinted with permission from Ref.[123]

consistent with the LSV results, in which selectivity of the electrocatalyst decreases toward NRR. The stability of AuHNCs-715 is evaluated for 60 h by conducting 5 consecutive cycles, each for 12 h, at −0.4 V (Figure 3.18c). The electrocatalyst could maintain the continuous NH_3 formation with a stable NH_3 yield rate and FE (Figure 3.18d). In addition, the SEM images before and after the durability test show the morphology of supported nanoparticles is reasonably maintained after 60 h of the CA test (Figure 3.19). Overall, these findings lead to empirical structure-activity trends for ammonia synthesis by an array of hollow Au–Ag nanocatalysts with tunable plasmonic properties.[123]

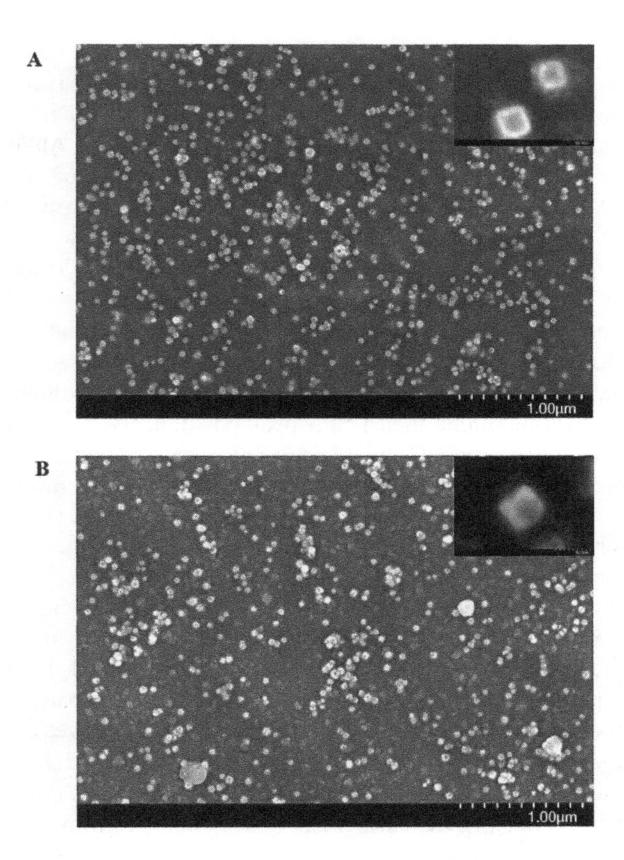

FIGURE 3.19 The SEM images (a) before and (b) after the durability test. The electrode after the test was thoroughly washed with DI water and dried at room temperature before taking the measurement. Reprinted with permission from Ref.[123]

3.6 THE ROLE OF OXIDATION OF SILVER IN BIMETALLIC GOLD-SILVER NANOCAGES ON THE ELECTROCATALYTIC ACTIVITY OF NRR

The electrocatalyst's chemical stability during electrochemical NRR is of paramount importance when considering the feasibility of electrochemical NRR for industrial applications. Therefore, it is essential to investigate the role of possible oxidation of an electrocatalyst under reaction conditions. In the case of Au–Ag nanocages, it is vital to discern the effect of oxidation of Ag on the selectivity and activity of the electrocatalyst toward NRR.[125] This also leads to comparing Au and Ag atoms' active sites in the electrocatalyst upon oxidation of Ag. Ag_2O–Au nanocages are synthesized through the facile oxidation process to mimic the possible oxidation of Au–Ag electrocatalyst at the presence of O_2 in the electrolyte during electrochemical NRR.

By oxygenating the bimetallic porous Au–Ag nanocages with various LSPR peak positions through purging the solution with pure oxygen gas, Ag is oxidized to form silver (I) oxide (Ag_2O) at room temperature, an extremely stable metal oxide semiconductor at ambient conditions. After oxygen treatment of the Au–Ag nanocages, the LSPR peak position slightly redshifts (e.g., 9 nm for Au–Ag-635), suggesting the formation of Ag_2O in the cavity and successful synthesis of Ag_2O–Au nanocages (Figure 3.20a). Furthermore, the SERS spectrum of Ag_2O shows distinct peaks at 460, 685, 817, 964, 1083 cm^{-1}, which are all attributed to the Ag-O vibrational modes (Figure 3.20b). These peaks with a slight shift are observed for Ag_2O–Au nanocages, while no pronounced SERS peak is observed for the Au–Ag nanocages, further confirming the formation of Ag_2O after oxygen treatment of Au–Ag nanocages (Figure 3.20b).[126] The size and density of pores in the walls of hollow Au–Ag nanocages are controlled by tuning their LSPR peak position. By increasing the amount of Au^{3+} ions added to the Ag–NCs template, Ag atoms are etched and replaced by Au atoms. This results in the LSPR peak position redshift from 635 nm to 715 nm, and the pore size at the walls and corners of the nanocages increases (Figures 3.20c and 3.20e). The pore size and morphology of nanocages did not change significantly after the O_2 treatment of Au–Ag nanocages (Figures 3.20d and 3.20f).

The detailed information regarding the LSPR peak positions before and after oxygen treatment of bimetallic Au–Ag nanocages and the Au and Ag content of nanoparticles which are determined by ICPES are provided in Table 3.4. Energy dispersive X-ray (EDX) spectroscopy was performed on a single nanocage particle to determine the shell composition (Figure 3.21 and Table 3.5).[127,128] EDX revealed the O (at. %) content of Au–Ag nanocages increases after O_2 treatment (Table 3.5). Since the areas are selected from the shell of nanocages to determine the surface composition, there is a discrepancy between EDX and ICPES values obtained from the bulk analysis. Before O_2 treatment of nanocages, the initial O content is attributed to the chemisorbed oxygen caused by the external –OH group or the water molecule adsorbed on the surface (Table 3.5).

The peaks observed at potentials around 0.5 V vs. RHE from CV measurements correspond to the reduction of silver (I) oxide in Ag_2O–Au-644 and Ag_2O–Au-719 nanocages. These reduction peaks are slightly larger than those using Au–Ag-635 and Au–Ag-715 nanocages, while the reduction peak for Au oxide at potential around 1.2 V vs. RHE remained relatively unchanged (Figure 3.22a)[123]. This observation suggests the effective conversion of Ag to Ag_2O after O_2 treatment of bimetallic Au–Ag nanocages and successful synthesis of Ag_2O–Au with various LSPR peak positions. This also informs that the O_2 treatment of Ag in bimetallic Au–Ag nanocages to form Ag_2O is more effective than the electrochemical oxidation of Ag.

The selectivity performance of Ag_2O–Au nanocages with various LSPR peak positions is evaluated toward NRR. The higher current density was achieved for both electrocatalysts in N_2-saturated electrolyte compared to Ar-saturated electrolyte within the potential window of −0.2 V to −0.5 V vs. RHE (Figure 3.22b). In addition, within this potential window, the selectivity decreases for both electrocatalysts (i.e., Ag_2O–Au-644 and Ag_2O–Au-719) after O_2 treatment of bimetallic Au–Ag nanocages (Table 3.6). This reveals the role of Ag in promoting the selectivity toward NRR. For instance, at the potential −0.4 V vs. RHE, the selectivity decreases from

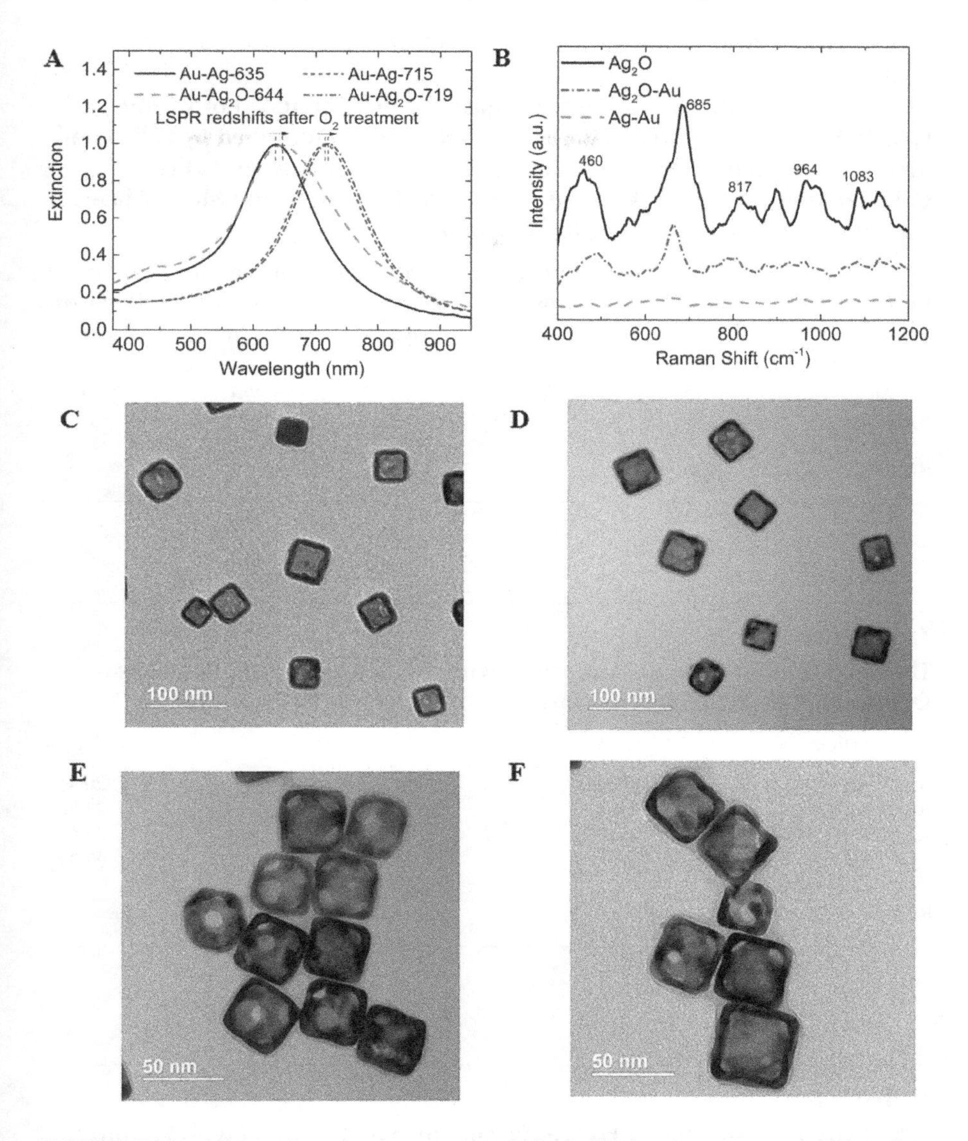

FIGURE 3.20 (a) UV-vis extinction spectra of bimetallic Au–Ag nanocages with various LSPR peak positions before and after O_2 treatment. Before O_2 treatment, the bimetallic Au–Ag nanoparticles have the LSPR peak positions at 635 nm and 715 nm while after O_2 treatment the LSPR redshifts to 644 nm and 719 nm, suggesting the formation of Ag_2O in the cavity (Ag_2O–Au). (b) SERS spectra of Ag_2O, Ag_2O–Au-685, and Ag–Au-670 nanoparticles. The TEM images of bimetallic Au–Ag nanocages with the LSPR peak position (c) at 635 nm before O_2 treatment (Au–Ag-635) and (d) at 644 nm after O_2 treatment (Ag_2O–Au 644). The TEM images of bimetallic Au–Ag nanocages with the LSPR peak position (e) at 715 nm before O_2 treatment (Au–Ag-715) and (f) at 719 nm after O_2 treatment (Ag_2O–Au-719). Reprinted with permission from Ref.[125]

TABLE 3.4
Au, Ag Concentrations, Au Content (mass % and Atomic %) of Various Hybrid Hollow Plasmonic Nanoparticles that are Determined by ICPES and their Corresponding LSPR Peak Position. Atomic Content (at. %) Is Calculated using Au and Ag's Concentrations Divided by the Molar Mass of Au (196.97 g mol^{-1}) and Ag (107.87 g mol^{-1})

Electrocatalyst	LSPR (nm)	Au Conc. (μg mL^{-1})	Ag Conc. (μg mL^{-1})	Au Content (mass %)	Au Content (atom %)
Ag–Au	635	1.2	2.44	33.0	21.2
Ag_2O–Au	644				
Ag–Au	715	3.91	3.45	53.1	38.3
Ag_2O–Au	719				

Reprinted with permission from Ref.[125]

TABLE 3.5
The Shell Composition of Nanocages with various LSPR Peak Positions Determined by EDX Spectroscopy

Nanoparticle	Ag (at. %)	Au (at. %)	O (at. %)
Ag–Au-635	73.63	5.96	20.41
Ag_2O–Au-644	65.82	5.68	28.50
Ag–Au-715	67.78	10.67	21.55
Ag_2O–Au-719	61.69	10.94	27.37

Reprinted with permission from Ref.[125]

TABLE 3.6
Selectivity Performance of Hybrid Plasmonic Nanoparticles with Various LSPR Peak Positions Toward NRR

	Potential (V vs. RHE)		
Electrocatalyst	−0.3	−0.4	−0.5
Ag–Au-635	52.3	48.7	45.1
Ag_2O–Au-644	46.4	36.7	9.1
Ag–Au-715	55.6	65.3	56.3
Ag_2O–Au-719	51.7	48.2	32.1

Reprinted with permission from Ref.[125]

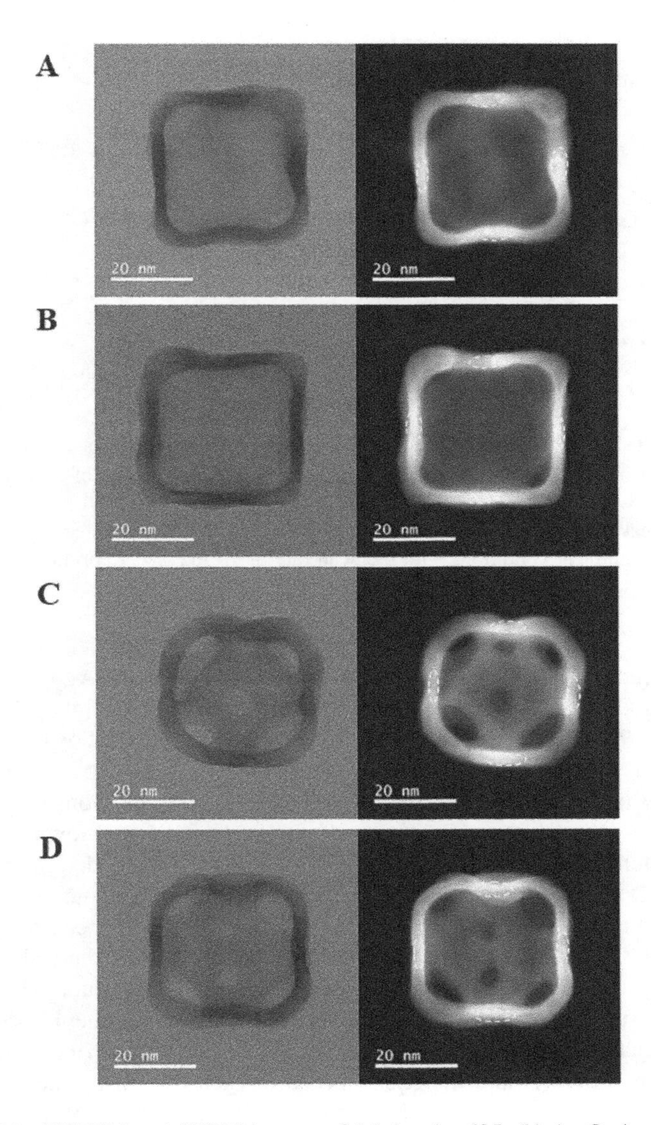

FIGURE 3.21 HRTEM and STEM images of (a) Ag–Au-635, (b) Ag_2O–Au-644, (c) Ag–Au-715, and (d) Ag_2O–Au-719 nanocages. Reprinted with permission from Ref.[125]

65.3% to 48.2% after O_2 treatment of Au–Ag-715 nanocages to make Ag_2O–Au-719. Suppose Au atoms were the only active sites for NRR in bimetallic Au–Ag nanocages. In that case, the selectivity should remain unchanged after O_2 treatment of Au–Ag nanocages, as Au is a stable metal catalyst even after oxygenation. Since Ag_2O is a p-type semiconductor, at the reduction potentials (e.g., −0.4 V vs. RHE) for NRR, it acts as a resistor rather than conductor, and therefore it is not an active site for NRR. However, Au atoms serve as primary active sites for NRR since for all

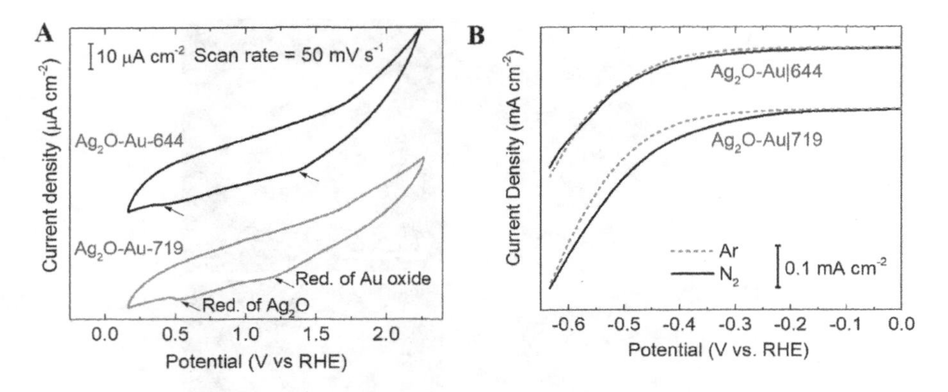

FIGURE 3.22 (a) Cyclic voltammograms (CV) of Ag_2O–Au nanocages with various LSPR peak positions in Ar-saturated 0.1 M LiOH (aq.) solution at a scan rate of 50 mV s^{-1}. The CV measurements were conducted in the RDE setup at room temperature. (b) LSV tests of Ag_2O–Au with LSPR peak positions at 644 nm and 719 nm in an Ar and N_2 saturated 0.5 M $LiClO_4$ (aq.) solution under ambient conditions with the scan rate of 10 mV s^{-1}. Reprinted with permission from Ref.[125]

potentials studied, higher selectivity has been achieved using Ag_2O–Au-719 compared to Ag_2O–Au-644 (Table 3.6). This is attributed to the higher Au content in Ag_2O–Au-719 compared to Ag_2O–Au-644, which were revealed by both ICPES and EDX analysis (Tables 3.4 and 3.5). It is important to note that as potential moves from −0.3 V to −0.5 V, the selectivity toward NRR decreases from 51.7% to 32.7% for Ag_2O–Au-719 and from 46.4% to 9.1% for Ag_2O–Au-644. Moving toward more negative potentials than −0.6 V, HER becomes dominant. To determine the Fermi level of Ag_2O and better understand its response to the applied potential, Ag_2O nanocubes are synthesized by O_2 treatment of AgNCs at the LSPR peak position of 445 nm. After O_2 treatment, the extinction spectrum dampens, and the LSPR peak position redshifts (11 nm), confirming the formation of Ag_2O nanocubes (Figure 3.23a). Furthermore, the O 1 s profile for Ag nanocubes after O_2 treatment is deconvoluted into two peaks centered at 531.5 eV and 532.6 eV, which are attributed to the lattice oxygen atoms of Ag_2O and the chemisorbed oxygen caused by the external –OH group or the water molecule adsorbed on the surface (Figure 3.23b).[129,130] The latter peak is also observed in the O 1 s profile for Ag nanocubes. The Mott–Schottky (MS) test determines the flat band potential (E_{fb}) and identifies the semiconductor type of Ag_2O (Figure 3.23c). The E_{fb} is determined to be 1.31 V vs. RHE. The negative slope of the MS curve confirms that Ag_2O is a p-type semiconductor. At potentials more negative than E_{fb} for a p-type semiconductor such as Ag_2O, a depletion region exists. At potentials more positive than E_{fb} for a p-type semiconductor, an accumulation region arises. In the depletion region for Ag_2O (the potential window for NRR is in the depletion region of Ag_2O), there are a few charge carriers available for charge transfer, and therefore electron transfer reactions happen very slowly. This behavior of Ag_2O is in line with the selectivity performance of Ag_2O–Au-644 and

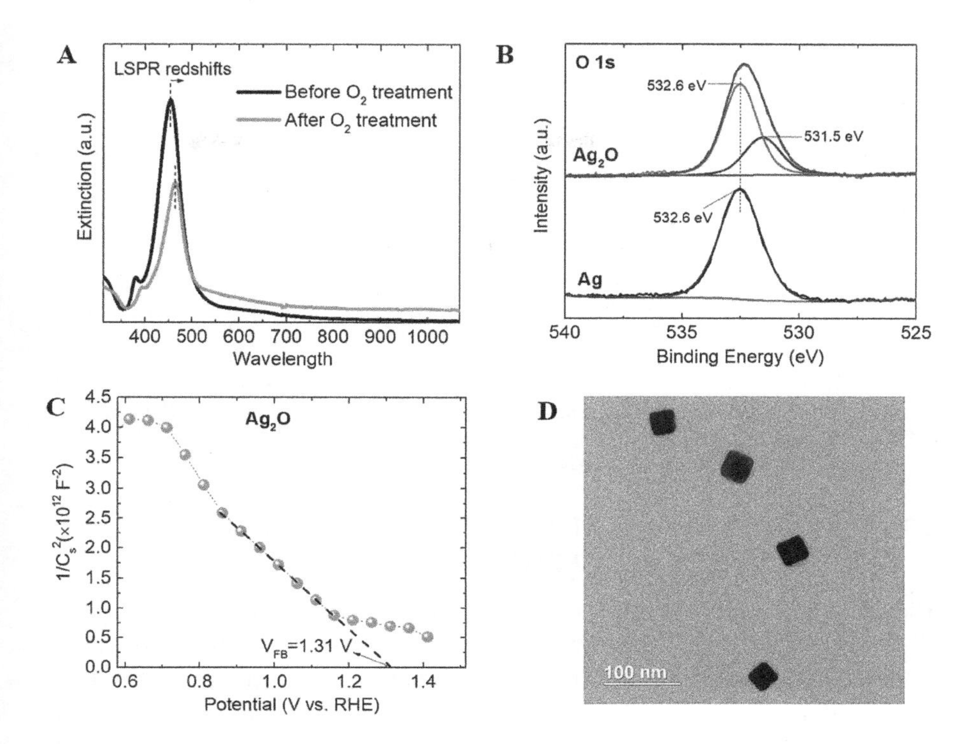

FIGURE 3.23 (a) UV-vis extinction spectra of AgNCs before and after O_2 treatment. (b) O 1 s profiles of Ag nanocubes before and after O_2 treatment. All spectra were shift corrected using a standard reference C1s, C–C peak at 284.8 eV. (c) Mott–Schottky plot of Ag_2O. A Mott–Schottky plot at a frequency of 1000 Hz was measured in 0.5 M $LiClO_4$ (aq.) solution under the dark condition. (d) The TEM image of solid Ag_2O nanocubes. Reprinted with permission from Ref.[125]

Ag_2O–Au-719 in which, by sweeping the potential to more negative potentials (more negative from E_{fb}), the selectivity toward NRR continuously decreases (Table 3.6). The TEM image of solid Ag_2O nanocubes reveals Ag nanocubes' morphology is maintained after the O_2 treatment of nanoparticles (Figure 3.23d). The ammonia yield rate and FE decrease from 1.58 µg $cm^{-2}\ h^{-1}$ and 16.4% to 0.98 µg $cm^{-2}\ h^{-1}$ and 11.4% after O_2 treatment of Au–Ag-635 (Figure 3.24a). Higher ammonia yield rate and FE are obtained using Ag_2O–Au-719 (2.14 µg $cm^{-2}\ h^{-1}$ and 23.4%) compared to Ag_2O–Au-644, consistent with the selectivity performance of these two electrocatalysts for NRR. Again, these results further confirm that although Ag acts as an active site for NRR, it has lower NRR activity than Au atoms. The same trend is observed at various applied potentials. The ammonia yield rate and FE decrease after O_2 treatment of bimetallic Au–Ag nanoparticles, with the highest ammonia yield rate FE, achieved at −0.4 V vs. RHE (Figure 3.24b). It should be emphasized the importance of an O_2-free electrolyte and careful synthesis and treatment of bimetallic Au–Ag nanocages to avoid the formation of Ag_2O in the cavity, which causes a significant reduction in the electrocatalytic NRR activity.

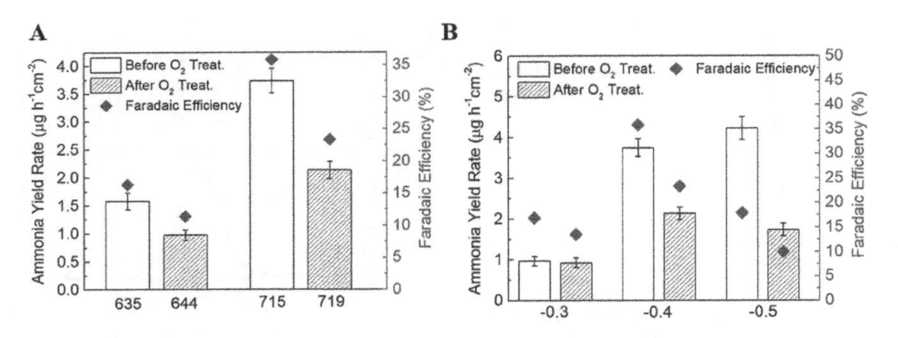

FIGURE 3.24 (a) Ammonia yield rate and FE for bimetallic Au–Ag nanocages before and after O_2 treatment with various LSPR peak positions at a potential of −0.4 V vs. RHE in 0.5 M $LiClO_4$ (aq.) solution. (b) Ammonia yield rate and FE at various applied potentials in a 0.5 M $LiClO_4$ (aq.) solution. Reprinted with permission from Ref.[125]

3.7 INCORPORATING THE TRANSITION METAL INTO PLASMONIC NANOCATALYSTS

The rate-determining step kinetics is accelerated on transition metal hydrides, where H^* species on the catalyst surface directly react with the dissolved nitrogen in the electrolyte to form N_2H_x.[131–133] In addition, a lithium-mediated strategy ($Li + \xrightarrow{e^-} Li \xrightarrow{N_2} Li_3N \xrightarrow{H^+} NH_3$) on the transition metal catalyst could enhance the electrochemical NRR activity at low temperatures due to lithium's unique ability to break the triple nitrogen bond, followed by hydrogenation to generate ammonia.[134,135] The optimization of the structure, morphology, and composition is critical to enhancing the rate of electro-reduction of N_2 to NH_3. It is desired to design a hybrid catalyst by incorporating catalytic transition metals into plasmonic nanoparticles with high NRR activity.

This new metal must correctly modify the nanoparticles' electronic structure by shifting the d-band center to ensure an adsorbed species' optimum binding energy on the catalyst surface.[136] It was demonstrated that the electrochemical NRR takes place on the surface of Pd catalyst through the formation of Pd hydrides (PdH_x) at low overpotentials, followed by surface hydrogenation reactions.[133] The formation of PdH_x could weaken Pd-H bonding (proton adsorption strength on the Pd surface), leading to enhanced binding of Pd surface with N_2 and N-containing adsorbates.

Since the reduction potential of Pd (~0.65 V vs. RHE) is higher than that of Ag (~0.5 V vs. RHE) but lower than that of Au (~1.2 V vs. RHE), Ag can only be replaced with Pd in the galvanic replacement process after the addition of Pd salt (K_2PdCl_4 (aq.)) in the bimetallic Au–Ag nanostructures template. The red shifting of the LSPR peak position of the trimetallic nanostructures indicates either the replacement of Ag with Pd or the growth of Pd on Au. The production energy efficiency (η_{PEE}) of the electrolysis system is calculated according to Equation 3.14:

$$\eta_{PEE}(\%) = \frac{\Delta G \textit{ for ammonia generation}\left(J\,mol^{-1}\right) \times \textit{ammonia generated}\left(mol\right)}{\int IVdt\left(J\right)} \times 100 \quad (3.14)$$

The free energy for NH_3 generation (ΔG) is 339 kJ mol^{-1} and $\int IVdt$ (J) is the electricity consumed in the process (energy input). Here, V (V) is the full cell potential and is calculated according to Equation 3.15:

$$V = (E_{anode} - E_{cathode}) + IR \tag{3.15}$$

The cathode potentials (NRR) were measured in the three-electrode set up for each condition.

The anode potential is estimated as the sum of the thermodynamic potential for water oxidation (1.23 V vs. RHE) and OER overpotential (~0.45 V on the Pt mesh).[137] The resistance (R) between the anode and the cathode in our setup should be measured and included in the full cell potential. The ECSA of the catalyst is determined from the charge associated with the reduction peak of Pd oxide (PdO) after double-layer correction and is normalized to the Pd loading on the working electrode and the charge density of 424 μC cm^{-2}.

Bimetallic porous Au–Ag nanocages with an LSPR peak position at 650 nm are synthesized by the galvanic replacement method by adding $HAuCl_4$ (aq.) solution to a solution of solid silver nanocubes (AgNCs) (Figures 3.25a and 3.25b). Trimetallic Au–Ag–Pd nanostructures are synthesized by adding K_2PdCl_4 (aq.) solution to the porous bimetallic Au–Ag nanocages dispersed in DI water with the LSPR peak at 650 nm. The volume of the Pd^{2+} precursor added to the Au–Ag template, and subsequent reduction of Pd^{2+} to Pd^0 (two Ag atoms are replaced with one palladium atom), is controlled by measuring the redshift in the LSPR peak positions of nanoparticles. As the amount of Pd salt solution increases, the amount of redshift and the LSPR bandwidth increases (Figure 3.25a). Excessive addition of Pd salt results in aggregation and collapsing of the nanocages. The LSPR peak of the resulting trimetallic Au–Ag–Pd redshifts from 650 nm to 850 nm; indicating the replacement of the remaining Ag atoms in the Au–Ag nanocages with Pd atoms (galvanic replacement) or the growth of Pd on Au (Figure 3.25). Due to the lattice mismatch between Pd (3.890 Å) and Au (4.079 Å), when Pd^{2+} precursor is added to the solution of Au–Ag nanocages, islands of palladium are formed on the Au surface. By further addition of the Pd salt solution, the islands are grown more, building a highly rough and porous surface (Figures 3.25c and 3.25d). Scanning transmission electron microscopy (STEM) and EDX spectroscopy are performed on a single trimetallic nanostructure to determine the structure, the elemental composition, and distribution of Au–Ag–Pd-750 and Au–Ag–Pd-850 nanoparticles (Figure 3.26). For Au–Ag–Pd-750, Pd islands are distributed in the Au–Ag nanocages structure, indicating the successful incorporation of Pd into bimetallic Au–Ag nanocages, and the synthesis of trimetallic Au–Ag–Pd nanoparticles (Figures 3.26a–f). By increasing the amount of Pd salt solution and red shifting the LSPR peak from 750 nm to 850 nm, a continuous porous layer of Pd is formed at the exterior surface of the nanoparticles (Figures 3.26g–l). In addition, ICPES measurements reveal that by adding the Pd salt solution into the Au–Ag nanocages, the Au content remains relatively unchanged while the Ag content decreases from 2.68 mg L^{-1} to 1.35 mg L^{-1} when the LSPR peak redshifts from 650 nm to 850 nm (Table 3.7). This indicates that the galvanic replacement of Ag atoms with Pd ions is one of the mechanisms for the reduction of Pd^{2+} to Pd^0 in which

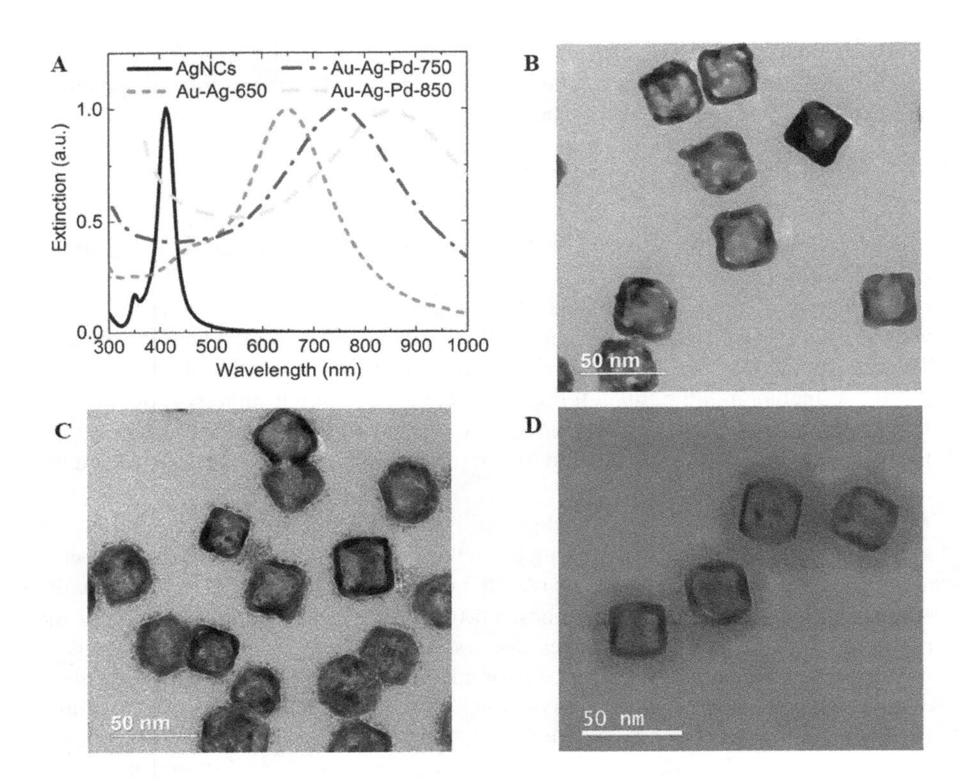

FIGURE 3.25 (a) UV-vis extinction spectra of silver nanocubes, bimetallic Au–Ag nanocages, and trimetallic Au–Ag–Pd nanoparticles with the LSPR peak positions at 750 nm and 850 nm. TEM images of (b) Au–Ag-650, (c) Au–Ag–Pd-750, and (d) Au–Ag–Pd-850. Silver nanocubes with the LSPR peak position at 412 nm are used as a template to synthesize various bimetallic and trimetallic nanoparticles. Reproduced with permission from Ref.[142]

the concentration of Pd increases to 1.23 mg L^{-1} and 2.77 mg L^{-1} as the LSPR peak redshifts to 750 nm and 850 nm. The stoichiometric balance between Ag and Pd (Pd^{2+} (aq.) + $2Ag^0$ (s) $\rightarrow$ Pd^0 (s) + $2Ag^+$ (aq.)) and the Pd and Ag contents in Table 3.7 reveal that the reduction of Pd^{2+} to Pd^0 is also accomplished through the layer by layer growth on Au atoms (Figure 3.26i).[16,138] X-ray photoelectron spectroscopy (XPS) results of Au–Ag–Pd-850 nanoparticles reveal a spin-orbit doublet for Au 4f at 84.0 eV and 87.7 eV, indicating that the zero-valence state of Au (Au^0) is preserved after the addition and reduction of the Pd^{2+} precursor in bimetallic Au–Ag nanocages (Figure 3.27a). The Ag 3d doublet energy peaks at 367.9 eV and 373.9 eV slightly shifts to lower binding energy (0.2 eV) compared to the Ag 3d doublet in bimetallic Au–Ag nanocages; this is attributed to the interaction and charge distribution of metallic Ag^0 and Pd precursors that occur after galvanic replacement (Figure 3.27b).[139] The Pd 3d spectrum is deconvoluted into two pairs of doublets. The doublet peaks at 335.8 eV, and 340.8 eV correspond to Pd at zero-valence state, suggesting the successful incorporation of Pd^0 in the bimetallic Au–Ag nanocages. The doublet peaks

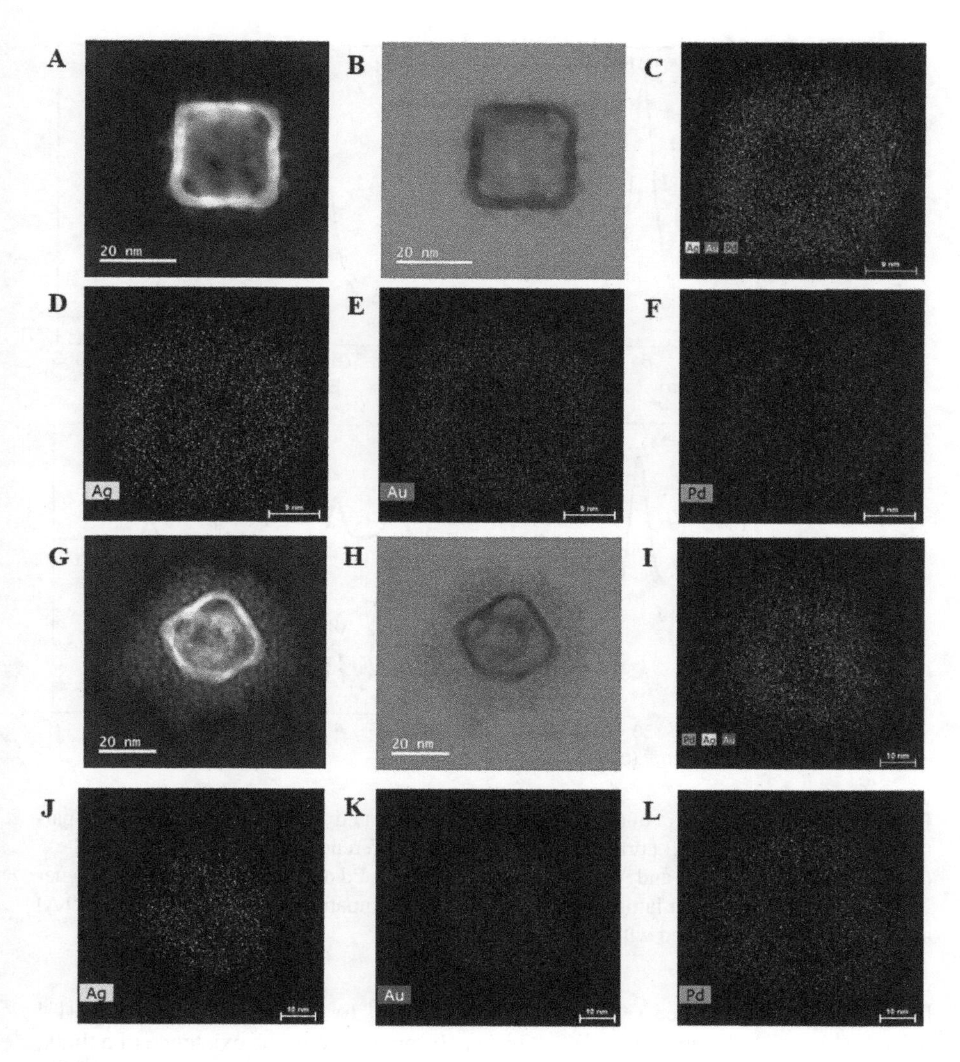

FIGURE 3.26 STEM (a), HRTEM (b), and EDX elemental mapping (c–f) of a representative single Au–Ag–Pd-750 nanoparticle. STEM (g), HRTEM (h), and EDX elemental mapping (i–l) of a representative single Au–Ag–Pd-850 nanoparticle. Reprinted with permission from Ref.[142]

at higher binding energies and with lower intensities than those of zero-valence Pd metal (337.1 eV and 342.5 eV) are attributed to the oxidized Pd states (Pd^{n+}) (Figure 3.27c).[140] The XRD patterns of Au–Ag–Pd and Au–Ag nanocages indicate the presence of metallic Pd, which is evident from the three distinct peaks at higher 2 angles (40.4°, 46.9°, 68.4°) in Au–Ag–Pd compared to those of the Au–Ag nanocages (38.3°, 44.6°, 64.7°) (Figure 3.27d).

The ECSAs of the nanoparticles are normalized to the Pd and Au loading on the working electrode, obtained from ICPES measurements (Table 3.7). Au–Ag–Pd-850

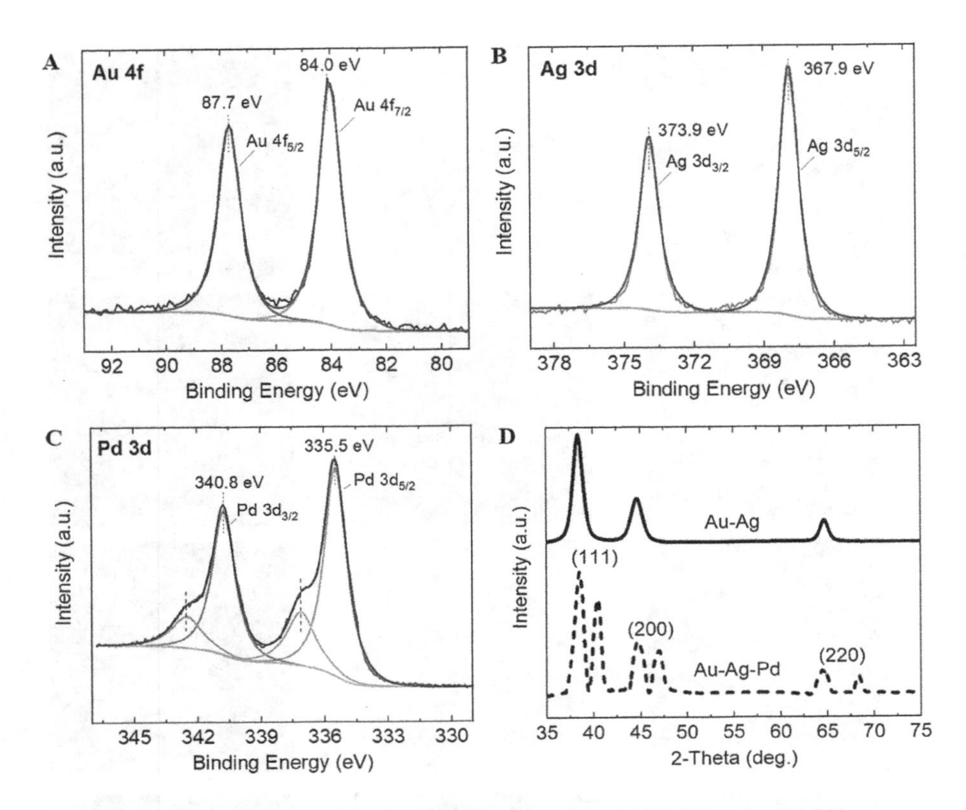

FIGURE 3.27 XPS spectra of (a) Au 4f, (b) Ag 3d, and (c) Pd 3d of Au–Ag–Pd-850 nanoparticles. All spectra were shift corrected using a standard reference C1s, C–C peak at 284.8 eV. (d) XRD pattern of Au–Ag and Au–Ag–Pd nanocages. The Pd diffraction peaks shift to higher 2θ angles due to the smaller lattice parameter (3.890 Å) compared to those of the Au (4.079 Å) and Ag (4.086 Å). Reprinted with permission from Ref.[142]

has a higher surface area (345.4 $m^2\ g^{-1}$) compared to the Au–Ag–Pd-750 (291.4 $m^2\ g^{-1}$) (Figure 3.28a and Table 3.7). This can be attributed to the existence of a thick, porous layer of Pd at the exterior surface of the Au–Ag–Pd-850 nanoparticles, which increases the ECSA for catalytic reactions. It is well known that rough surfaces are catalytically more active than smooth surfaces, as atoms present on rough surfaces are more thermodynamically active.[13,16] The $ECSA_{Pd}$ obtained using our trimetallic Au–Ag–Pd-850 nanoparticles is approximately five times higher than the $ECSA_{Pd}$ of the commercial Pd/C catalysts.[122,133] Although the reduction peak area of PdO is remarkably higher than that of Ag oxide, due to the comparable reduction potentials of these two metal oxides, the reduction peak area of Ag oxide is subtracted from the actual reduction peak area centered at around 0.5 V vs. RHE to determine the $ECSA_{Pd}$ (Figure 3.28a). The observed negative shift in the electrochemical reduction of PdO (0.1 V) in our trimetallic nanostructures, compared to the pure Pd, is attributed to the slight alloying of Pd and Ag at their interface during the galvanic replacement reaction.[141] The $ECSAs_{Au}$ of trimetallic Au–Ag–Pd nanoparticles is comparable (0.93) to

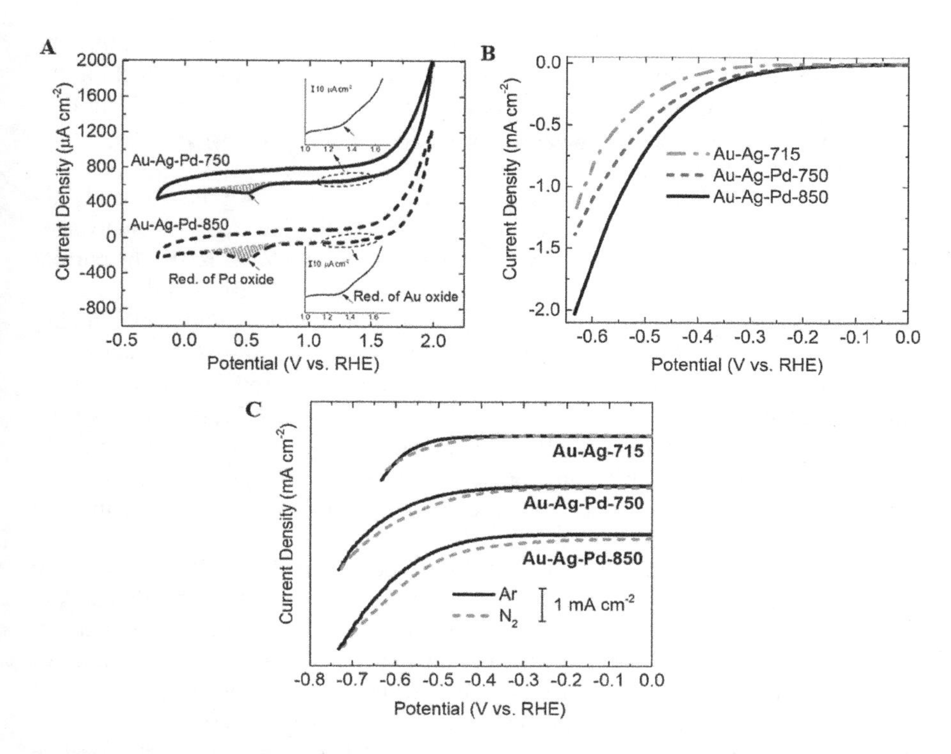

FIGURE 3.28 (a) Cyclic voltammograms (CV) of Au–Ag–Pd-750 and Au–Ag–Pd-850 in Ar-saturated 0.1 M LiOH (aq.) at a scan rate of 50 mV s^{-1}. The CV measurements were conducted in the RDE setup at the stagnant electrode condition at room temperature. (b) LSV of various catalysts in N_2-saturated 0.5 $LiClO_4$ (aq.). (c) LSV tests of Au–Ag–Pd-750 and Au–Ag–Pd-850 in an Ar- and N_2-saturated 0.5 M $LiClO_4$ (aq.) under ambient conditions with the scan rate of 10 mV s^{-1}. Reprinted with permission from Ref.[142]

TABLE 3.7

Au, Ag, and Pd Concentrations of Nanoparticles are Determined by ICPES. The electrochemical Surface Areas (ECSA) of Nanoparticles are Determined based on the Reduction Peak of Au Oxide and PdO during CV Measurements in Ar-saturated 0.1 M LiOH Solution at a Scan Rate of 50 mV s^{-1}

Catalyst	Ag Conc. (μg mL^{-1})	Au Conc. (μg mL^{-1})	Pd Conc. (μg mL^{-1})	Au Content (at. %)	Pd Content (at. %)	$ECSA_{Au}$ (m^2 g^{-1})	$ECSA_{Pd}$ (m^2 g^{-1})
Au–Ag-650	2.68	1.32	NA	21.25	NA	23.3	NA
Au–Ag–Pd-750	1.88	1.31	1.23	18.66	32.43	20.8	291.4
Au–Ag–Pd-850	1.35	1.29	2.77	14.52	57.72	22.2	345.4

Reprinted with permission from Ref.[142]

the ECSAsAu of bimetallic Au–Ag nanocages (Table 3.7). This indicates that the incorporation of Pd in bimetallic Au–Ag nanocages does not result in blocking the Au active sites for electrochemical NRR and confirms the preservation of the hollow structure in trimetallic nanoparticles where Au, Ag, and Pd can act as catalytically active centers for the reaction.

A remarkably higher current density is achieved using Au–Ag–Pd-750 and Au–Ag–Pd-850 nanoparticles compared to the bimetallic Au–Ag-715 nanocages (Figure 3.28b). For instance, at an applied potential of −0.3 V vs. RHE, the current density increases 4.8 and 3.7 times for Au–Ag–Pd-850 and Au–Ag–Pd-750 compared to the Au–Ag-715. This observation is consistent with significantly higher ECSAPd compared to the $ECSA_{Au}$ (Table 3.7) in trimetallic Au–Ag–Pd nanoparticles, which provides more active sites for nitrogen adsorption and reduction. Although achieving high current density at low overpotentials does not necessarily indicate the high selectivity and activity of an electrocatalyst toward NRR, it is the primary step toward the high production rate that is mandatory for reaching the overarching goal of commercializing electrochemical NRR for sustainable ammonia production.

The higher current density was achieved in N_2-saturated electrolyte compared, to the Ar-saturated electrolyte, for both electrocatalysts (i.e., Au–Ag–Pd-750 and Au–Ag–Pd-850) within the wide potential window (Figure 3.28c). Unlike the bimetallic Au–Ag nanocages demonstrated in previous chapters, the selectivity performance of electrocatalysts improves at low overpotentials (0 to −0.3 V vs. RHE) toward NRR after the incorporation of Pd into the bimetallic Au–Ag nanocages (Figure 3.28c). In addition, both electrocatalysts are selective toward NRR until the high negative potential of −0.75 V vs. RHE, even with an enhanced current density that is achieved using trimetallic nanostructures. This expands the selectivity of an electrocatalyst toward NRR in negative potentials by 0.15 V over that of Au–Ag nanocages. Moving toward more negative potentials than −0.75 V, a HER becomes dominant. The potential window of −0.3 to −0.5 V vs. RHE, is a compromise of obtaining high current density and selectivity. It is important to note that achieving a high current density for NRR is crucial to decrease the capital cost, leading to the commercialization of electrochemical NRR. At all three applied potentials, the selectivity is higher for Au–Ag–Pd-850 than for the Au–Ag–Pd-750; this is attributed to the higher Pd content and lower Ag content of Au–Ag–Pd-850 nanoparticles. Even though the selectivity is not substantially higher in trimetallic Au–Ag–Pd nanostructures than in the bimetallic Au–Ag nanocages, due to the enhanced current density obtained (I_{N2}) using Au–Ag–Pd nanoparticles, the corresponding current density toward NRR (I_{N2}–I_{Ar}) is much more pronounced after incorporation of Pd into Au–Ag nanocages. The analysis of the sample solution after CA measurements revealed that the highest ammonia yield rate (13.74 μg cm^{-2} h^{-1}) is obtained at −0.4 V corresponding to an FE of 44.11% while the highest FE (48.94%) is achieved at −0.3 V (ammonia yield rate = 5.80 μg cm^{-2} h^{-1}) using Au–Ag–Pd-850 (Figure 3.29a and 3.29b). The higher ammonia yield rate but lower FE at −0.4 V, compared with −0.3 V, is attributed to the compromise between increasing current density and competitive selectivity toward HER rather than NRR. For all potentials tested, both the NH_3 yield rate and the FE are higher when Au–Ag–Pd-850 is used than when Au–Ag–Pd-750 is used. This result suggests

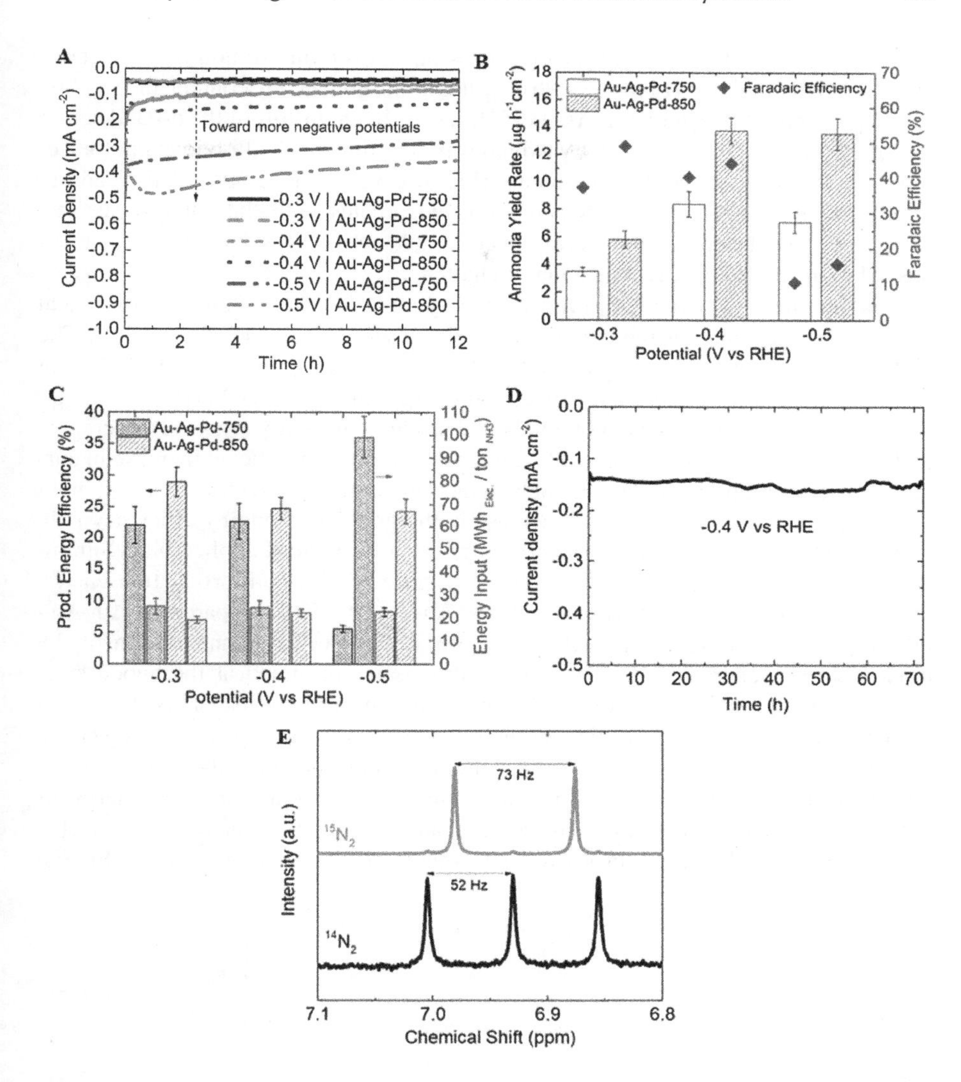

FIGURE 3.29 (a) CA results of Au–Ag-Pd-750 and Au–Ag–Pd-850 nanocatalysts at a series of potentials. (b) Ammonia yield rate and FE at various potentials in 0.5 M $LiClO_4$ (aq.) solution using Au–Ag–Pd-750 and Au–Ag–Pd-850 nanocatalysts. (c) Production energy efficiency and energy input at various applied potentials using Au–Ag–Pd-750 and Au–Ag–Pd-850 nanocatalysts. The intense bars (black bars show production energy efficiency and blue bars show energy input) represent Au–Ag–Pd-750 and the medium bars represent Au–Ag–Pd-850. (d) CA test for the stability of the Au–Ag–Pd-850 at −0.4 V vs. RHE in 0.5 M $LiClO_4$ (aq.) solution. (e) ^1H NMR spectra of samples after electrochemical $^{15}N_2$ ($^{14}N_2$) reduction reaction at −0.4 V vs. RHE for 4 h in 0.5 M $LiClO_4$ (aq.) solution. Reprinted with permission from Ref.[142]

that Pd content and $ECSA_{Pd}$ of nanoparticles have a role in promoting the electrocatalytic NRR activity (Figure 3.29b). In addition, greater Pd content in Au–Ag–Pd-850 compared to that of Au–Ag–Pd-750 results in an upshift of the d-band center ($E - E_f$) from −4.45 eV to −3.73 eV, which was determined via ultraviolet photoelectron spectroscopy (UPS) measurements (Figures 3.30a and 3.30b). This leads to improved binding strength of N-containing adsorbates with the catalyst surface, which is a crucial step for engineering selective and active NRR catalysts in an aqueous solution where selectivity is a major challenge.

Production energy efficiency (%) and energy input (MWh $_{Elec.}$/ton $_{NH3}$) are critical parameters in evaluating the performance of the N_2 electrolysis system[105,142]. The highest production energy efficiency (28.9%) is achieved using Au–Ag–Pd-850 at −0.3 V, which corresponds to the electrical energy input of 19.1 MWh $_{Elec.}$/ton $_{NH3}$. Moving toward more negative potentials, the production energy efficiency decreases to 24.8% at −0.4 V and 8.3% at −0.5 V, mainly due to the significant increase in current density at more negative potentials, which increases the electrical energy input (Figure 3.29c). It is noted that the increase in the production energy efficiency is in line with the decrease in the electrical energy input for various applied bias. Similar to the electrocatalytic NRR activity results (Figure 3.29b), both production energy efficiency and energy input deteriorate for Au–Ag–Pd-750 compared to Au–Ag–Pd-850, in various applied potentials (Figure 3.29c). A significant portion of the input electrical energy (>75%) for all conditions is consumed at the anode side, where oxygen evolution reaction (OER) takes place. Alternative use of organic-based electrolytes (e.g., glycerol) in the anodic half-reaction could remarkably lower the electricity consumption.[143] It is worth mentioning that the state-of-the-art thermochemical process (Haber–Bosch) for ammonia synthesis consumes the energy of 7.8 MWh $_{Elec.}$/ton $_{NH3}$ (based on natural gas), and the target production energy efficiency of advanced research projects agency-energy (ARPA-E) program (REFUEL)

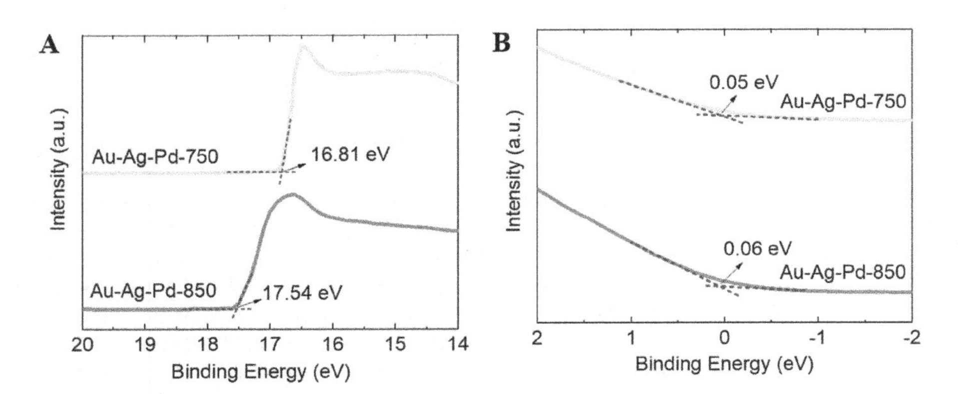

FIGURE 3.30 UPS spectra of trimetallic Au–Ag–Pd nanoparticles. (a) The secondary electron cut-off ($E_{cut\text{-}off}$) and (b) the Fermi edge (E_{FE}) are measured with He_I (21.22 eV) source radiation. The work function (∅) is calculated by 21.21 eV − ($E_{cut\text{-}off} - E_{FE}$). The ∅ of Au–Ag–Pd-750 and Au–Ag–Pd-850 nanoparticles are determined to be 4.45 and 3.73 eV. Reprinted with permission from Ref.[142]

for electrochemical fuel production is greater than 60%[36,144]. The electrocatalyst could maintain the NH_3 yield rate (13.62 μg cm^{-2} h^{-1}) and FE (43.5%) over 72 h, which is very close to the NRR activity of the catalyst in the initial 12 h test (Figures 3.29d and 3.31a). This performance corresponds to the turnover frequency (TOF) of 59 h^{-1} as per active Pd, Au, and Ag sites. The turnover number (TON) per active Pd, Au, and Ag sites in trimetallic nanoparticles is determined by dividing the total moles of ammonia produced within 72 h to the total moles of active sites (e.g., Pd, Au, and Ag) according to Equation 3.16:

$$TON = \frac{N_{ammonia}}{N_{Pd} + N_{Au} + N_{Ag}} \tag{3.16}$$

The concentration of each element in the nanoparticle is provided in Table 3.7. By dividing the concentration of each atom by the molecular weight, the concentration with the unit of mol L^{-1} is obtained. The volume of the nanoparticle drop-casted on the substrate is 300 μL.

$$N_{Pd} = \frac{2.77\, mg\, L^{-1}}{106.42 \times 1000\, mg\, mol^{-1}} \times 300 \times 10^{-6} L \cong 7.81 \times 10^{-9} mol \tag{3.17}$$

$$N_{Au} = \frac{1.29\, mg\, L^{-1}}{196.97 \times 1000\, mg\, mol^{-1}} \times 300 \times 10^{-6} L \cong 1.965 \times 10^{-9} mol \tag{3.18}$$

$$N_{Ag} = \frac{1.35\, mg\, L^{-1}}{107.87 \times 1000\, mg\, mol^{-1}} \times 300 \times 10^{-6} L \cong 3.755 \times 10^{-9} mol \tag{3.19}$$

The total mole of active sites is:

$$\begin{aligned} N_{total} = N_{Pd} + N_{Au} + N_{Ag} &= 7.81 \times 10^{-9} + 1.965 \times 10^{-9} + 3.755 \times 10^{-9} \\ &= 1.353 \times 10^{-8}\, mol \end{aligned} \tag{3.20}$$

It is important to note that this calculation is valid with the assumption that all Pd, Au, Ag atoms are fully converted by aqua regia (HCl + HNO_3 (3:1) vol. %) to their ionic form for the ICPES measurement.

The total amount of ammonia produced using Au–Ag–Pd-850 nanoparticles over 72 h is:

$$N_{ammonia} = \frac{13.62\, \mu g\, cm^{-2}\, h^{-1} \times 72\, h \times 1\, cm^2}{17 \times 10^6\, \mu g\, mol^{-1}} = 5.77 \times 10^{-5}\, mol\, of\, NH_3 \tag{3.21}$$

The TON within 72 h is:

$$TON = \frac{5.77 \times 10^{-5}\, mol\, of\, NH_3}{1.353 \times 10^{-8}\, mol\, of\, active\, sites} = 4264.6 \tag{3.22}$$

The TOF is defined as TON per unit time, which is given by:

$$TOF = \frac{4264.6}{72} \sim 59\,h^{-1} \tag{3.23}$$

Extensive control experiments must be carried out to understand the source of ammonia formation in the electrochemical system.[145–147] Control experiments (i.e., Ar, N_2 at OCV, N_2 with no catalyst) yield remarkably smaller amounts of NH_3 (Ar: 7.3% of N_2, N_2 at OCV: 6.1% of N_2, N_2 with no catalyst: 2.5% of N_2) (Figure 3.31b). The amount of ammonia produced in an isotopic labeling experiment using $^{15}N_2$ gas after a 4 h electrolysis test is close to that of $^{14}N_2$ (73.6 μM, 91% of $^{14}N_2$) (Figure 3.31c). Furthermore, the doublet and triplet couplings of $^{15}N_2$ and $^{14}N_2$ obtained from ^1H NMR measurement confirms that the supplied N_2 is the major source of ammonia formation in the system (Figure 3.29e). The amounts of ammonia measured using ^1H

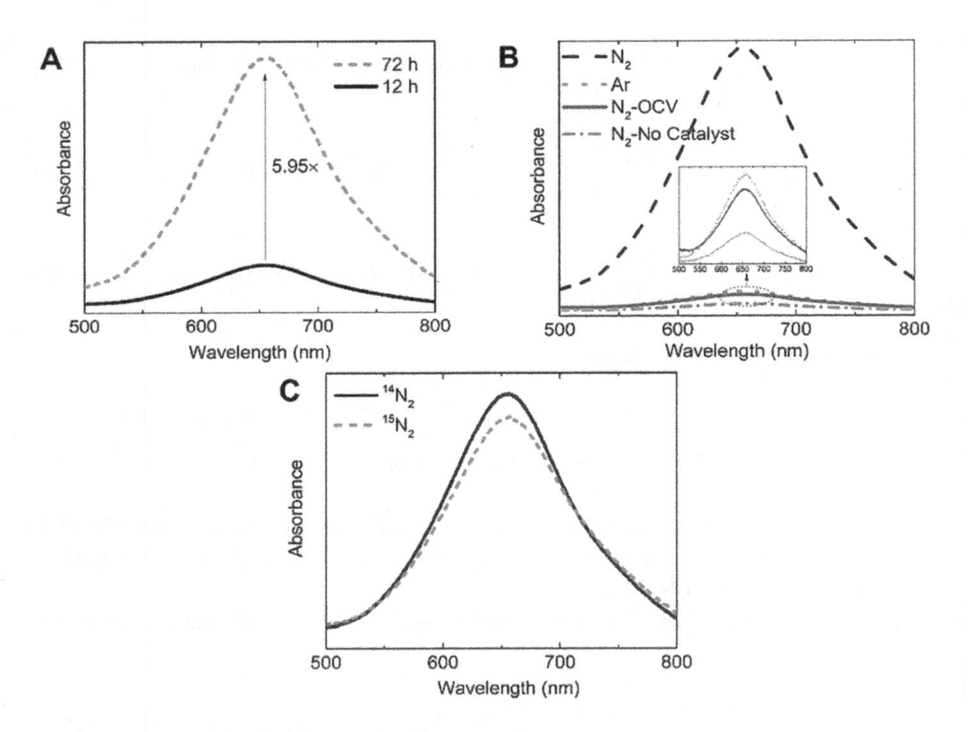

FIGURE 3.31 (a) UV-vis absorption spectra of 0.5 M $LiClO_4$ (aq.) solution after electrolysis at −0.4 V vs. RHE for 12 h using Au–Ag–Pd-850 nanoparticles obtained with indophenol blue method in various operating conditions (i.e., N_2, Ar, N_2 at OCV, N_2 with no catalyst). (b) UV-vis absorption spectra of 0.5 M $LiClO_4$ (aq.) solution after electrochemical NRR with $^{14}N_2$ and $^{15}N_2$ at −0.4 V vs. RHE for 4 h. (c) UV-vis absorption spectrum of 0.5 M $LiClO_4$ (aq.) solution after 72 h stability test at −0.4 V vs. RHE. Reprinted with permission from Ref.[142]

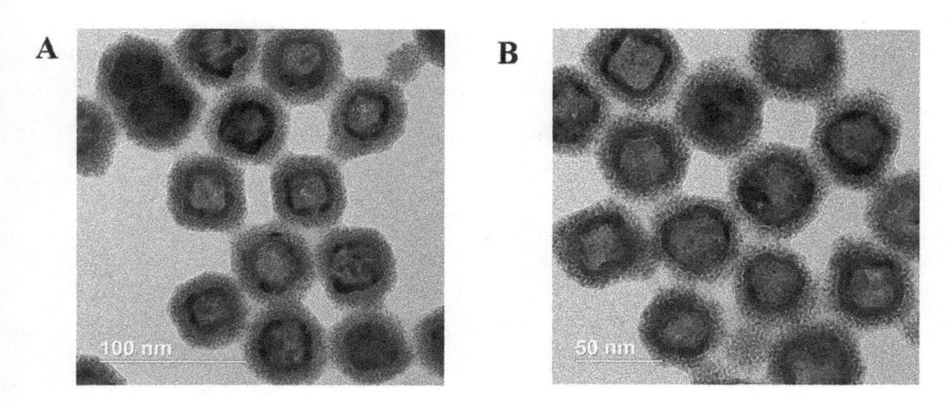

FIGURE 3.32 The TEM images (a) before and (b) after the electrocatalytic NRR stability test (72 h at −0.4 V vs. RHE in 0.5 M $LiClO_4$ (aq.) solution). Reprinted with permission from Ref.[142]

NMR are similar to the indophenol method (71.9 μM for $^{15}N_2$ and 78.4 μM for $^{14}N_2$). The TEM images after the stability test show the minor morphology change of nanoparticles after 72 h of the CA test (Figures 3.32a and 3.32b). Overall, it is essential to engineer the morphology and composition of nanocatalysts to improve the electrocatalytic NRR activity.

4 Electrochemical Reactor Design

4.1 INTRODUCTION

The electrochemical cell configuration is of paramount importance for conducting electrochemical conversion of N_2 to NH_3. Electrochemical designs are typically categorized as the single-chamber cell, double chamber cell (H-type cell), the membrane electrode assembly (MEA)-type cell. In 1968, Van Temelen et al. performed electrocatalytic N_2 conversion to NH_3 in a single compartment cell.[148] Various cell designs can be employed for electrochemical ammonia synthesis, which is distinguished by apparatuses, electrolytes, catalysts, and reaction parameters.[149] This chapter briefly discusses each electrochemical configuration for NRR, and the benefits and drawbacks of each design will be highlighted.

4.2 SINGLE CHAMBER CELL

Single chamber cells can utilize solid electrolytes (e.g., solid oxides) with a good proton or ion conductivity in an H_2 atmosphere. The solid electrolytes operating at high temperatures and atmospheric pressure. In this system, the H_2 is converted into H^+ at the anode, and the high H^+ flux is obtained at high temperatures (i.e., 570 °C) and then transferred to the cathode through the $SrCe_{0.95}Yb_{0.05}O_3$ solid electrolyte to drive the N_2 reduction reaction (Figure 4.1a).[150] This design's major drawbacks are high energy requirement for ammonia synthesis and NH_3 decomposition at high temperatures. NRR can also be achieved in molten-salt-based systems (e.g., 0.5 NaOH+0.5 M KOH electrolytes) at temperatures as high as 200 °C (Figure 4.1b).[90,151] Liquid-phase electrolytes in the single compartment cell have also received attention, enabling the production of ammonia under ambient conditions. The liquid electrolytes can be acid (e.g., HCl, H_2SO_4), alkaline (e.g., KOH, NaOH), and neutral solution (e.g., $LiClO_4$). A report demonstrated a high-pressure single chamber cell, where an H_2SO_4 solution is introduced into the methanol/$LiClO_4$ liquid electrolyte to provide a proton source (Figure 4.1c). The system operates at room temperature and high pressure of 50 bar to produce ammonia.[152] Single chamber cells can be adopted at room temperature and atmospheric pressure to synthesize NH_3. Careful attention should be paid to the choice of electrolyte's pH as each electrolyte has its advantages and disadvantages. For instance, acidic electrolytes might help provide a proton source for NRR; however, this acidic condition may also facilitate the competing hydrogen evolution reaction.[133,153] Ionic liquids (ILs) have been utilized to enhance ammonia yield, and the Faradaic efficiency in electrocatalytic NRR. ILs have shown higher N_2 solubility, nonvolatility, and robustness (Figure 4.1d).[154] Even though single chamber cells have been widely used for electrocatalytic NRR, they have inherent

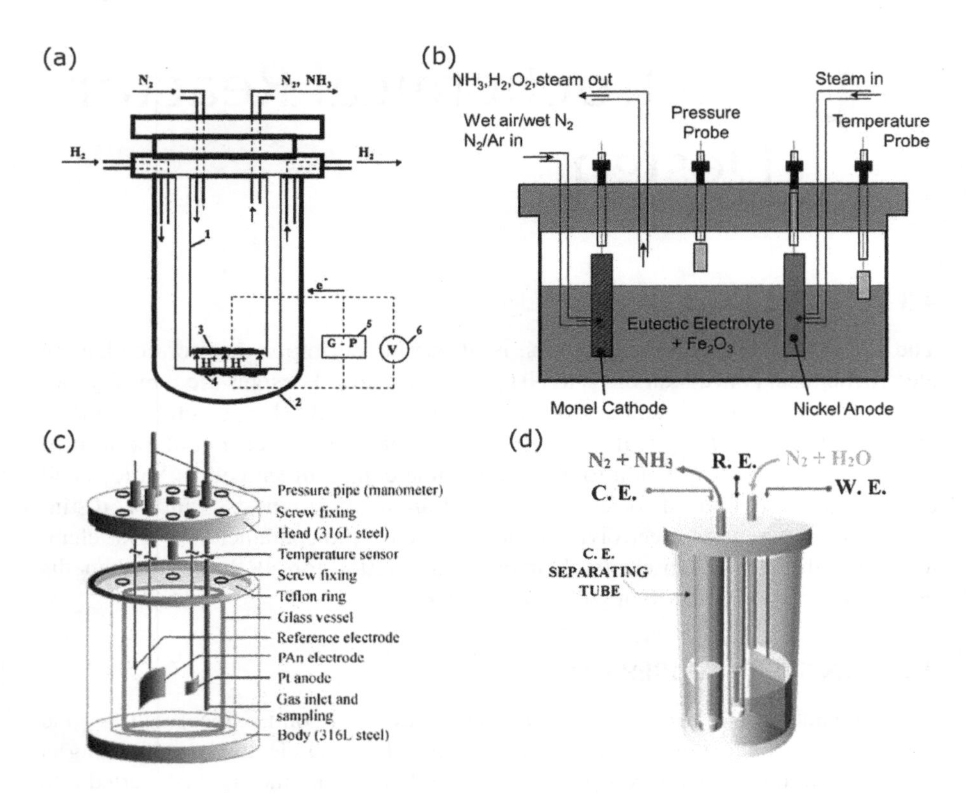

FIGURE 4.1 (a) Schematic diagram of a solid electrolyte single chamber cell for N_2 conversion to NH_3. Reproduced with permission from Ref.[150] (b) The electrochemical reactor for NH_3 synthesis from N_2 and steam in molten NaOH-KOH electrolyte with nano-Fe_2O_3 suspension. Reproduced with permission from Ref.[90] (c) High-pressure single chamber cell for electrocatalytic NRR, which operates at room temperature and 50-bar pressure using methanol/ $LiClO_4$ liquid electrolytes and H_2SO_4 as a proton source. Reproduced with permission from Ref.[152] (d) Schematic of the NRR electrochemical cell, where C.E., R.E., and W.E. represents counter electrode, reference electrode, and the working electrode, respectively. Reproduced with permission from Ref.[154]

limitations, including the mono experiment conditions for the anode and the cathode (e.g., operating pressure and temperature) and the possible diffusion of ammonia products to the anode, leading the oxidation of ammonia. This may limit the widespread adoption of a single-chamber cell for electrochemical ammonia synthesis.

4.3 TWO COMPARTMENT CELLS (H-TYPE CELLS)

In single-cell systems, both oxidation and reduction reactions occur in the same cell. The produced ammonia is susceptible to oxidation at the anode in this system. Separating the cathode and anode into two divided chambers (referred to as H-cell) is an effective way to address the deficiencies of single-chamber cells and improve the catalytic performance of NRR. In H-cells two compartments are separated by a proton exchange membrane (PEM) or an anion exchange membrane (AEM). A catalyst-coated substrate is used as the working electrode on the cathode side, while silver/silver chloride (Ag/AgCl) is the standard choice for the reference electrode. Pt mesh or foil and graphite rod are often used as the counter electrode (Figure 4.2a). As the reaction proceeds, the produced protons or hydroxide ions are transported across the PEM or AEM to balance the redox reactions under ambient catalytic conditions. The majority of the produced NH_3 is in the form of NH_4^+ in aqueous solutions, which cannot pass through the membranes, avoiding the oxidation and loss of products as occurs in single-chamber cells. Therefore, it is critical to select suitable membranes to minimize the NH_3/NH_4^+ products' crossover and measure the total NH_3 concentration in electrolytes from both anodic and cathodic chambers since a portion of products may still transfer through the membrane to the anode chamber. In addition, as membrane prevents species crossover, the electrolytes in two compartments can be different. This offers an opportunity to optimize redox reaction environments. For instance, in two compartments cell, a bi-electrolyte system in which LiCl/ethylenediamine (EDA) was selected as the cathodic electrolyte, and H_2SO_4 solution were chosen as the anodic electrolyte (Figure 4.2b).[155] This allowed a remarkable improvement in the Faradaic efficiency (FE) of 17.2% compared to the case, where 2-propanol was used as the electrolyte with an FE < 1%.[156] This is due to the EDA electrolyte's synergistic effect, which suppressed hydrogen evolution reaction (HER) at the cathodic compartment, and the H_2SO_4 electrolyte provides sufficient protons.

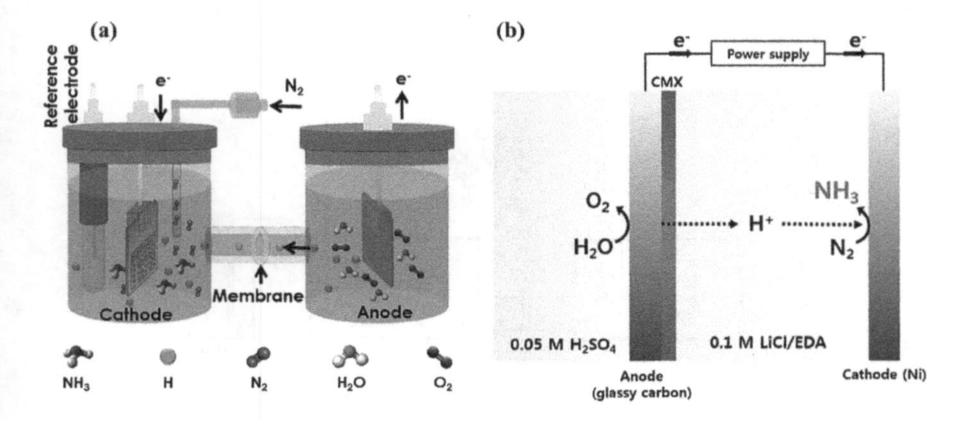

FIGURE 4.2 (a) Schematic diagram of an electrochemical H-cell for NRR. There are two compartments (anodic and cathodic) that are separated by a proton exchange membrane. Protons produced at the anode are transported across the membrane to the cathode where NRR occurs. (b) Schematic diagram of a double chamber cell with 0.1 m LiCl/EDA filled in the cathode chamber and 0.05 m H_2SO_4 filled in the anode chamber. A cation exchange membrane separates the two chambers. Reproduced with permission from Ref.[155]

4.4 MEMBRANE ELECTRODE ASSEMBLY (MEA)-TYPE CELL

The MEA contains a catalyst layer, a membrane, a gas diffusion layer (GDL), and current collectors (Figure 4.3a). Depending upon the reaction conditions, the membrane can be proton-conducting (e.g., acidic or neutral conditions) or hydroxide-conducting layers (e.g., basic environment). For instance, electrocatalytic NRR was conducted in an anion exchange membrane (AEM) cell with an electrode made of γ-Fe_2O_3 nanoparticles on porous carbon paper using 0.5 M KOH as the electrolyte to suppress HER (Figure 4.3b).[157] In the alkaline environment, N_2 directly reacts with water to produce ammonia, and hydroxide ions are transported across the AEM from the cathodic part to the anodic part, where they are oxidized to produce water according to the following reactions:

$$N_2 + 6H_2O + 6e^- \leftrightarrow 2NH_3 + 6OH^-, \quad E^o_{RHE} = 0.059\,V \tag{4.1}$$

$$6OH^- \leftrightarrow \frac{3}{2}O_2 + 3H_2O + 6e^-, E^o_{RHE} = 1.23\,V \tag{4.2}$$

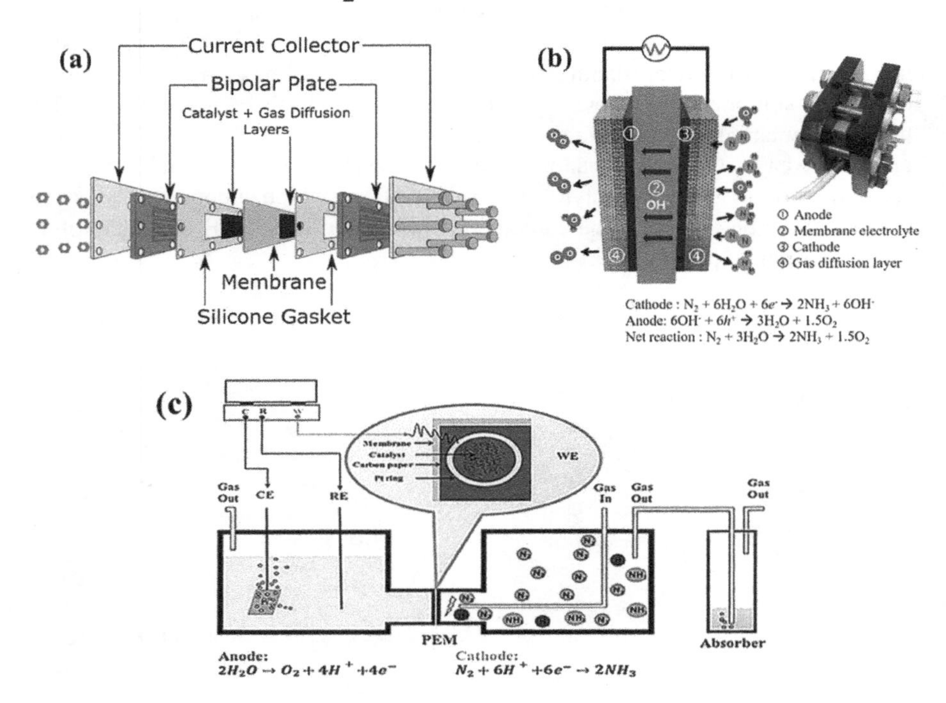

FIGURE 4.3 (a) Schematic illustration of the MEA-type cell for gas-phase electrochemical ammonia synthesis. The cell consists of a membrane, catalyst layers, gas diffusion layers, and current collectors. Reproduced with permission from Ref.[122] (b) Schematic of electrochemical synthesis of NH_3 in an anion-exchange-membrane-based electrolyzer. A photographic image of the actual device is shown in the inset. In this cell, AEM is used to facilitate the transport of hydroxide ions across the membrane. Reproduced with permission from Ref.[157] (c) Schematic view of the electrocatalytic flow reactor for ammonia synthesis, with the cathode cell operating under electrolyte-less conditions (gas phase). PEM indicates proton exchange membrane. Reproduced with permission from Ref.[84]

In the MEA-type cell, the redox reactions take place in the gas-phase conditions, where N_2 and H_2 or water vapor can be directly fed into the system, where the gaseous reactant, catalyst, and membrane meet.[158] This also allows overcoming the transport limitation due to the low solubility of N_2 in the aqueous electrolyte. The semi gas-phase system was demonstrated with half the cell in the liquid phase and the other half in the gas phase.[84] In this electrochemical setup, N_2 gas is directly flowing over the catalyst (Ru/carbon nanotubes (CNT)) without electrolytes, and the anode chamber is operated in $KHCO_3$ electrolyte to produce protons. The PEM-type membrane separates the chambers. This design enables the relatively easy separation of ammonia products from the gas stream.

Electrochemical systems can be directly integrated with light sources, leading to the development of the photoelectrochemical system. A typical photoelectrochemical system contains a light source, a reaction cell, and a three-electrode system with a catalyst-coated substrate responsive to the incident light as the working electrode, reference electrode, and counter electrode.[159,160] The photoelectrochemical reduction of N_2 occurs at the working electrode's surface (photocathode), and the water oxidation reaction occurs at the counter electrode. The reaction cells used in photoelectrochemical systems can be either single chamber cells[160,161] or H-cells.[162,163]

4.5 COMPARISON OF LIQUID- AND GAS-PHASE SYSTEMS FOR ELECTROCHEMICAL AMMONIA SYNTHESIS

Cost-effective production of ammonia via electrochemical NRR hinges on N_2 electrolysis at high current densities with suitable selectivity and activity. Gas-phase and liquid-phase electrochemical cells are primary cell configurations for ammonia synthesis under ambient conditions. While the gas-phase cell has lower ohmic losses and higher energy efficiency, the liquid-phase cell achieved higher selectivity and Faradaic efficiency, attributed to the presence of concentrated N_2 molecules dissolved in an aqueous electrolyte and the hydration effects. This section highlights the importance of design and optimization of cell configuration in addition to the modification of the catalyst to achieve high-performance N_2 electrolysis for ammonia synthesis. For small-scale and distributed ammonia synthesis reactors, an integrated system by coupling a photovoltaic (PV) cell to an electrochemical cell or a single photoelectrochemical device can be developed for solar-fuel based applications.[164–168] One of the challenges in electrochemical ammonia synthesis is finding catalysts that have a suitable activity for breaking N_2 triple bond in aqueous media under ambient conditions.[95–97,99] Improving the design of electrocatalysts, electrolytes, and electrochemical cells is required to overcome the selectivity and activity barrier in electrochemical NRR.[80,169] Prior studies on electrochemical NRR have focused on designing new electrocatalysts at low operating current densities (typically <200 $\mu A\ cm^{-2}$) with an enhanced N_2 selectivity in aqueous solutions or ionic liquids.[111,112,114,121,123,142,149,154,170,171] While nonaqueous solvents have been shown to suppress the HER and to improve the NRR Faradaic efficiency; they suffer from poor electrolyte conductivity and low energy efficiency.[109,172,173] To realize the commercialization of the electrochemical NRR process, N_2 electrolysis must be achieved at high current densities with high selectivity and activity. In addition to the design and optimization of electrocatalysts and electrolytes, another path to developing commercial NRR processes is designing electrochemical cells with low ohmic resistance. This translates to the facilitation of reactants' interactions with the electrode surface. Fuel cell-type electrochemical cells that consists of a MEA and GDLs have been shown to enable gas-phase NH_3 synthesis from N_2 and H_2.[122,158] It has been demonstrated that the kinetics of the rate-determining step in NRR ($N_2^* \rightarrow N_2H^*$) is accelerated on transition metal hydrides, where H^* species on the catalyst surface directly react with the dissolved nitrogen in the electrolyte to form N_2H^*.[132,133,174] The presence of Li^+ cations in the electrolyte enhances the electrochemical NRR due to the strong interaction of Li^+–N_2 and the ability of Li^+ to suppress the HER process by limiting the access of water molecules to the electrode surface.[85,116,175]

N_2 electrolysis for ammonia synthesis was studied using porous bimetallic Pd–Ag nanoparticles as an electrocatalyst in both a fuel cell type electrochemical cell (gas-phase) and H-cell (liquid-phase) at current densities above 1 mA cm^{-2} under ambient conditions.[176] Reactants in the liquid phase are N_2 and H_2O in a 0.5 M $LiClO_4$ (aq.) electrolyte, and gaseous N_2 and H_2 are reacted at the cathode and anode in the gas-phase system. A recent study found that increasing the Pd content (up to a Pd content of 57.7% at.) in porous trimetallic Au–Ag–Pd nanoparticles increases the electrocatalytic NRR activity.[142] Synthesizing porous bimetallic Pd–Ag nanoparticles that allow the Pd content to be increased to 80% while maintaining the nanostructure's porosity is hypothesized to increase the NRR activity.

Porous bimetallic Pd–Ag nanoparticles are synthesized by adding K_2PdCl_4 (aq.) solution to solid silver nanocubes (AgNCs) dispersed in DI water that has a localized surface plasmon resonance (LSPR) peak position at 412 nm (Figures 4.4a and 4.4b).[16] As the amount of Pd^{2+} precursor increases, the LSPR peak position redshifts, and the peak bandwidth increases (Figure 4.4a). PdNCs show only a background spectrum formed by d to d electron transitions (an interband transition spectrum). The redshift of the LSPR peak is due to the change in the shape of AgNCs as well as the change in the dielectric constant of the surrounding medium, resulting from the replacement of the Ag atoms in the AgNCs with Pd atoms (galvanic replacement) or the growth of Pd on Ag (Figure 4.4b). Due to the lattice mismatch between Pd (3.890$\mathring{A}$) and Ag (4.086$\mathring{A}$), when the Pd^{2+} precursor is added to the solution of AgNCs, islands of palladium are formed on the Ag surface. Further addition of Pd salt solution causes the islands to grow and create a rough and porous surface layer (Figure 4.4c and 4.4d). High-resolution transmission electron microscopy (HRTEM) shows lattice spacings of 0.232 nm and 0.224 nm, which correspond to the (111) lattice plane of the face-centered cubic (FCC) structure Ag and Pd (Figure 4.4c).

Scanning transmission electron microscopy (STEM) and energy dispersive X-ray (EDX) spectroscopy determine the structure, the elemental composition, and the distribution of Pd–Ag nanoparticles. The reduction of Pd^{2+} to Pd^o is accomplished through two different mechanisms. The galvanic replacement process leads to the formation of a hollow Pd–Ag nanostructure. The second mechanism is an island-growth mode that creates a continuous porous layer of Pd on Ag at the exterior surface (Figure 4.4e–g). The resulting nanoparticles achieved a Pd content of 81.2%.

The electrochemical surface area of Pd ($ECSA_{Pd}$) in the Pd–Ag nanoparticles is determined to be 277.5 m^2 g^{-1}, which is approximately four times higher than that of commercial Pd/C catalysts.[176] In a wide potential window ($-0.7 < E < 0$) from LSV

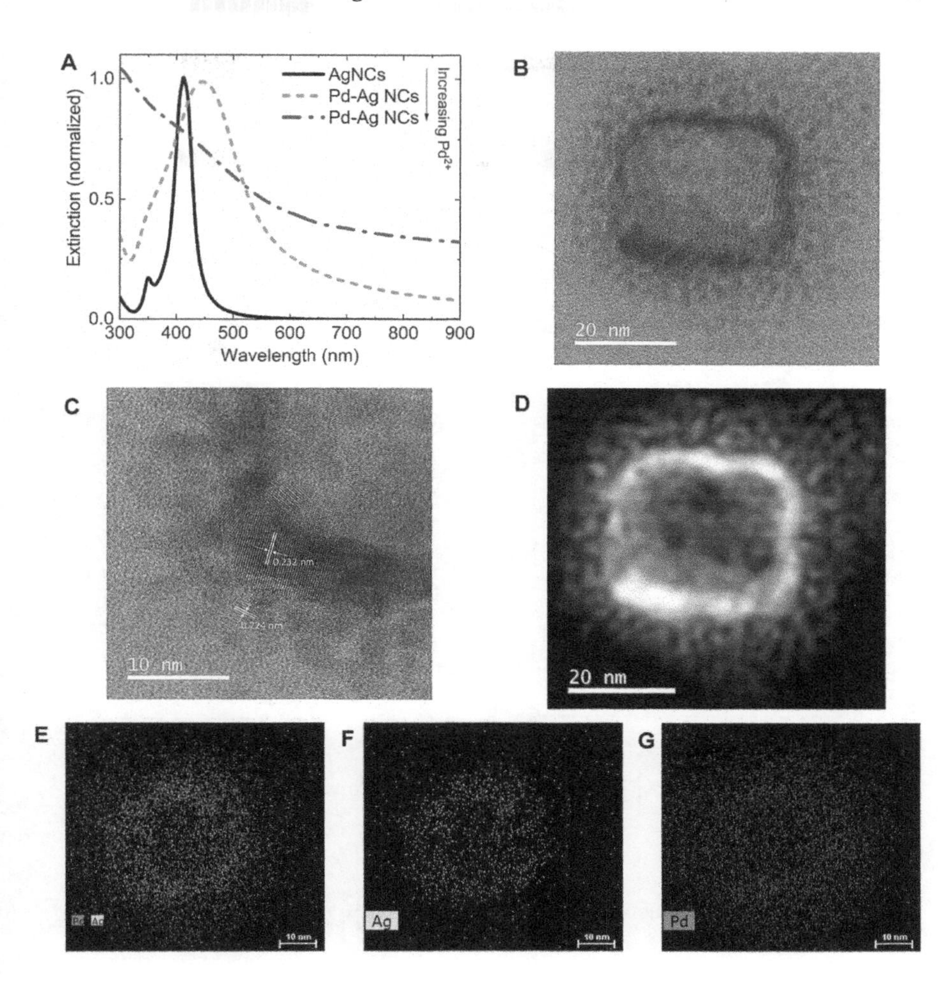

FIGURE 4.4 (a) UV--vis extinction spectra of silver nanocubes and bimetallic Pd – Ag nanocages with various amounts of Pd salt added. The LSPR peak of AgNCs shifts to red and becomes broader as the amount of Pd salt increases. TEM images of (b) Bimetallic Pd – Ag nanocages, (c) HRTEM image of Pd – Ag nanocages, (d) STEM, and (**e–g**) EDX elemental mapping of a representative single bimetallic Pd – Ag nanoparticle. Silver nanocubes with the LSPR peak position at 412 nm are used as a template to synthesize various bimetallic nanoparticles. Reprinted with permission from Ref.[176] Copyright 2020, American Chemical Society.

tests, a higher current density at a given potential is achieved in the N_2-saturated electrolyte (Figure 4.5a). At high negative potentials where HER becomes dominant, the N_2-saturated electrolyte has a higher current density than an Ar-saturated electrolyte, indicating that electrochemical NRR at higher current densities (in the order of milliamperes per centimeter squared) is feasible. CA tests are performed at a series of applied potentials using Pd–Ag nanoparticles to determine the ammonia yield rate and FE (Figures 4.5b and 4.5c). As the potential becomes more negative, the ammonia yield rate increases, reaching a maximum at −0.6 V (45.6 ± 3.7 $\mu g\ cm^{-2}\ h^{-1}$),

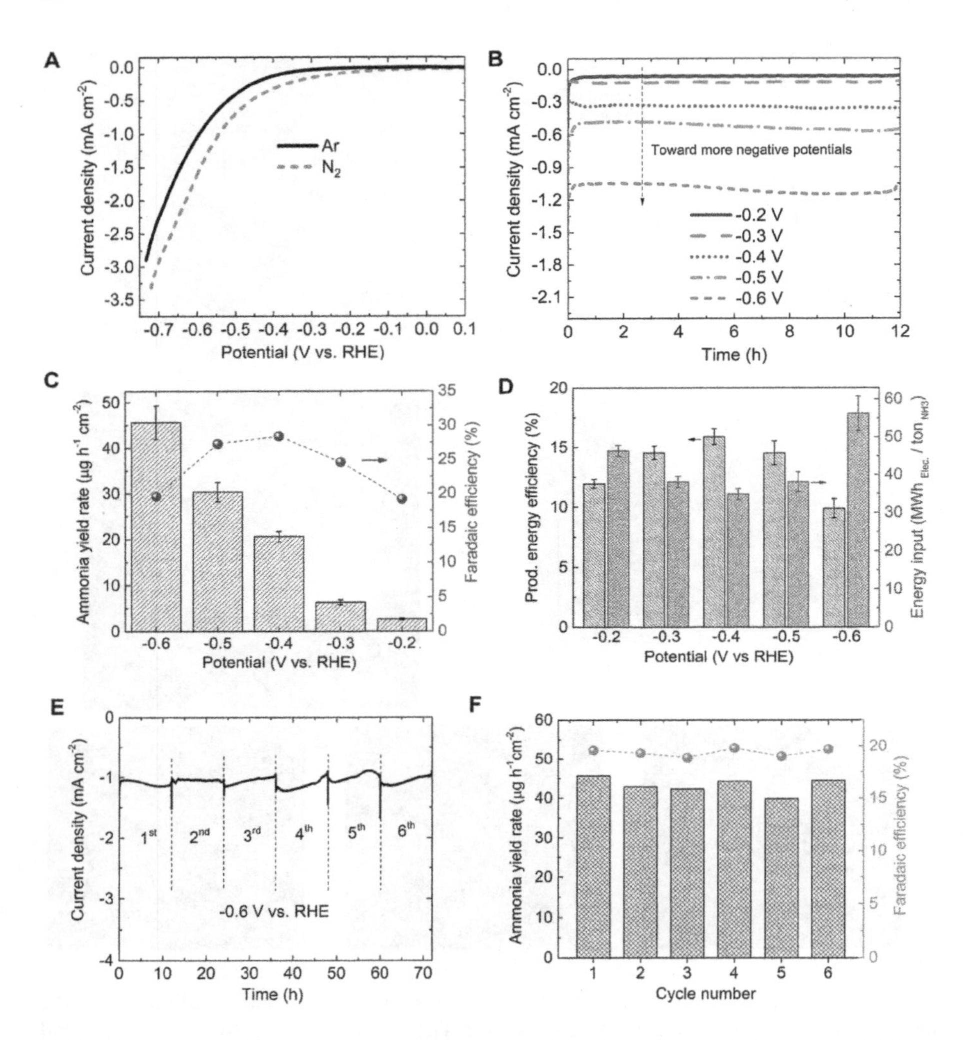

FIGURE 4.5 (a) LSV tests of Pd – Ag nanoparticles in an Ar- and N_2-saturated 0.5 M $LiClO_4$ (aq.) under ambient conditions with a scan rate of 10 mV s^{-1}. (b) CA results of Pd – Ag nanoparticles at a series of applied potentials. (c) Ammonia yield rate and FE at various applied potentials in 0.5 M $LiClO_4$ (aq.) solution using Pd – Ag nanoparticles. (d) Production energy efficiency and energy input at various applied potentials using Pd – Ag nanoparticles. (e) CA tests for the stability of Pd – Ag nanoparticles at −0.6 V vs. RHE in a 0.5 M $LiClO_4$ (aq.) solution. For each cycle, a CA test was carried out at−0.6 V vs. RHE for 12 h. (e) Cycling stability results of the ammonia yield rate and FE on Pd – Ag nanoparticles. Reprinted with permission from Ref.[176] Copyright 2020, American Chemical Society.

which corresponds to an FE of 19.6%. The NRR activity of different Pd-based catalysts is presented in Table 4.1. The FE at −0.6 V was lower than other applied potentials due to the compromise between increasing current density and competitive selectivity toward HER rather than NRR. The work function (Φ) of Pd–Ag nanoparticles is 4.54 eV, which is measured via ultraviolet photoelectron spectroscopy (UPS).[176] The downshift in the d-band center ($E - E_f$) of Pd–Ag nanoparticles (−4.54 eV) compared to that of trimetallic Au–Ag–Pd nanoparticles (−3.73 eV) in the previous report resulted in lower selectivity of N-containing adsorbates on the catalyst surface and therefore lower FE for Pd–Ag nanoparticles compared to the trimetallic Au–Ag–Pd nanoparticles.[142] However, due to the enhanced $ECSA_{Pd}$ and higher Pd content (81.2% at.), higher current densities and ammonia yield rates were observed using Pd–Ag nanoparticles. A production energy efficiency of 9.9% is achieved at −0.6 V, which corresponds to the electrical energy input of 56.2 MWh per ton of ammonia (Figure 4.5d). Most input electrical energy (>70%) is consumed at the anode side, where the oxygen evolution reaction (OER) takes place, which is added to additional ohmic losses (e.g., electrode and electrolyte resistances) in the liquid-phase system. The OER has a high thermodynamic potential (1.23 V) and sluggish reaction kinetics, resulting in a total cell potential higher than 2.0 V with the oxygen gas as a product that may be discharged and wasted. A CA test is performed at −0.6 V for 72 h by conducting six consecutive cycles, each for 12 h, to evaluate Pd–Ag nanoparticles' stability (Figure 4.5e). The electrocatalyst could maintain continuous NH_3 formation with an average NH_3 yield rate of 43.23 μg cm^{-2} h^{-1} and FE of 19.4% (Figure 4.5f). The turnover frequency (TOF) is 147 h^{-1} as per active Pd and Ag sites.[176] Before and after the stability test, SEM images reveal that the nanoparticles are still anchored to the substrate and are not washed away in the long-term experiment under stirring and applied bias (Figures 4.6a and 4.6b). TEM images after the stability test show minor morphology changes of the nanoparticles after 72 h of the CA test (Figures 4.6c and 4.6d). A monolayer of nanoparticles on a Si substrate is prepared via the Langmuir-Blodgett (LB) technique to evaluate the mechanical stability of hollow nanoparticles (Figure 4.7). LB technique was used to prepare a uniform monolayer of Pd–Ag nanoparticles on the substrate for the atomic force microscopy (AFM) measurement.[177] In this technique, Pd–Ag nanoparticles were dispersed in the ethanol:chloroform (2.5:1 vol. %) solvent, which is immiscible with the denser sublayer liquid (i.e., water) filling the LB trough. The surface tension of water forces the volatile solvent to form a uniform layer. As the volatile solvent evaporates, the nanoparticles will arrange into a monolayer on the water's surface. The mechanical barrier separates the dark-colored nanoparticles (left) from the unused water surface (right). The interparticle separation distance between the nanoparticles can be controlled by moving the mechanical barrier to decrease the trough area (Figure 4.7a–c). Pd–Ag nanoparticles in the volatile solvent (2 mL) are sprayed on the water surface, and after 30 min the surface pressure-surface area curve (isotherm) is obtained by decreasing the area at which the particles are dispersed (the left side of the mechanical barrier) and measuring the surface pressure at each surface area. The surface pressure was measured with a Wilhelmy plate attached to a D1L-75 model pressure sensor. Three distinct areas (i.e., gas phase, condensed liquid phase, solid phase) were observed depending upon the surface pressure and the interparticle distance (Figure 4.7d). In the gaseous state, the nanoparticles have a significant average separation like gases in a closed container. This phase is highly compressible, and changes in an area barely affect the surface pressure. The area at which we observe the gas to the liquid phase transition is determined by the relative strengths of the

TABLE 4.1
Summary of Studies on Electrochemical Ammonia Synthesis using Pd-based Nanoparticles (NPs) under Ambient Conditions. Reprinted with Permission from Ref.[176] Copyright 2020, American Chemical Society

Catalyst/Electrolyte	Applied Potential (V vs. RHE)	NH_3 Yield Rate (μg h^{-1} cm^{-2} or μg h^{-1} mg^{-1})	FE (%)	Ref.
Pd/C NPs | 0.1 M phosphate buffer solution (PBS)	0.1	4.5	8.2	185
PdCu NPs | 0.5 M LiCl	−0.1	35.7	11.5	186
PdRu NPs | 0.1 M KOH	−0.2	37.23	1.85	187
PdRu NPs | 0.1 M HCl	−0.1	25.92	1.53	188
AuPdP NPs | 0.1 M Na_2SO_4	−0.3	18.78	15.44	189
PdCuIr NPs | 0.1 M Na_2SO_4	−0.3	13.43	5.29	190
PdCu alloy | 1 M KOH	−0.25	39.9	<1	191
PdAg NPs | 0.5 M $LiClO_4$	−0.6	45.6	19.2	176

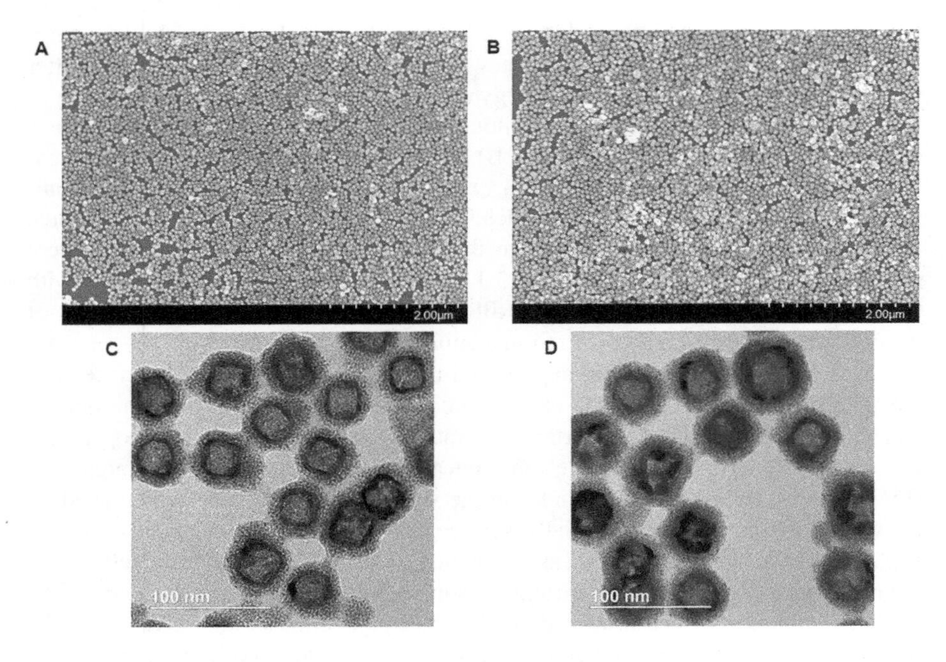

FIGURE 4.6 SEM images of nanoparticles (a) before and (b) after, and TEM images of nanoparticles (c) before and (d) after the electrocatalytic NRR stability test (6 consecutive cycles each for 12 h (total 72 h) at−0.6 V vs. RHE in 0.5 M $LiClO_4$ (aq.) solution. Reproduced with permission from Ref.[176] Copyright 2020, American Chemical Society.

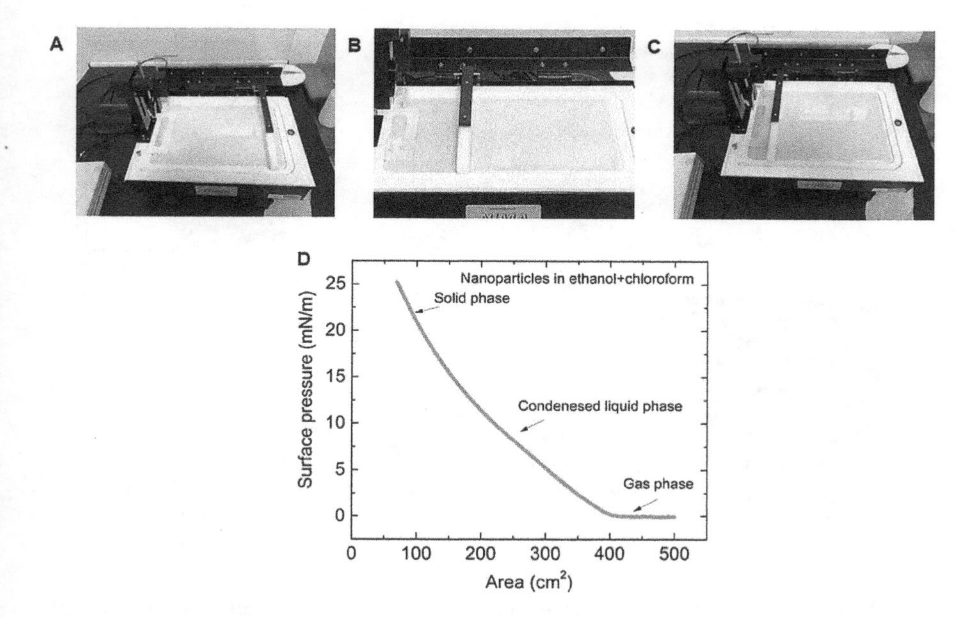

FIGURE 4.7 (a–c) Langmuir-Blodgett trough after dispersing Pd–Ag nanoparticles on top of the water sublayer. As the mechanical barrier moves to the left (decreasing the trough area), the surface pressure increases, resulting in more condensed nanoparticles with smaller separation distances. (d) Langmuir-Blodgett isotherm (surface pressure vs. area) for Pd–Ag nanoparticles measured on the top of water sublayers. Reproduced with permission from Ref.[176] Copyright 2020, American Chemical Society.

particle–particle and particle–water surface interactions. Eventually, the liquid phase changes to a solid phase, which consists of a dense and highly ordered film. The LB film was transferred to the silicon substrate by the vertical dipping method (5 mm/s vertical dipping speed), with the pressure kept constant during the deposition.

As hollow nanoparticles are more fragile than their solid counterparts and are susceptible to failure and collapse during catalysis, nanoparticles' stiffness after the long-term stability test is measured via AFM (Figure 4.8). The stiffness of Pd–Ag nanoparticles is measured to be primarily within the range of 1 to 6 GPa, which is comparable to the stiffness of hollow Pd nanoparticles of similar size and structure in prior studies.[178,179] By increasing the reaction temperature, the NRR activity increases, which is attributed to the faster mass transport rate of reactants at higher temperatures (Figures 4.9a and 4.9b). The overall activation energy of the electrochemical NRR on Pd–Ag nanoparticles is determined to be 73.6 kJ mol^{-1} (Figure 4.9c).

Similar experiments were conducted in a nonaqueous electrolyte (0.5 M $LiClO_4$ in tetrahydrofuran (THF) solution and ethanol (1% vol.)) (Figure 4.10a, b). A low current density (~8 $\mu A\ cm^{-2}$) is achieved at a high applied potential (−3.0 V vs. Ag/AgCl), resulting in energy efficiency lower than that of aqueous solution at a comparable FE

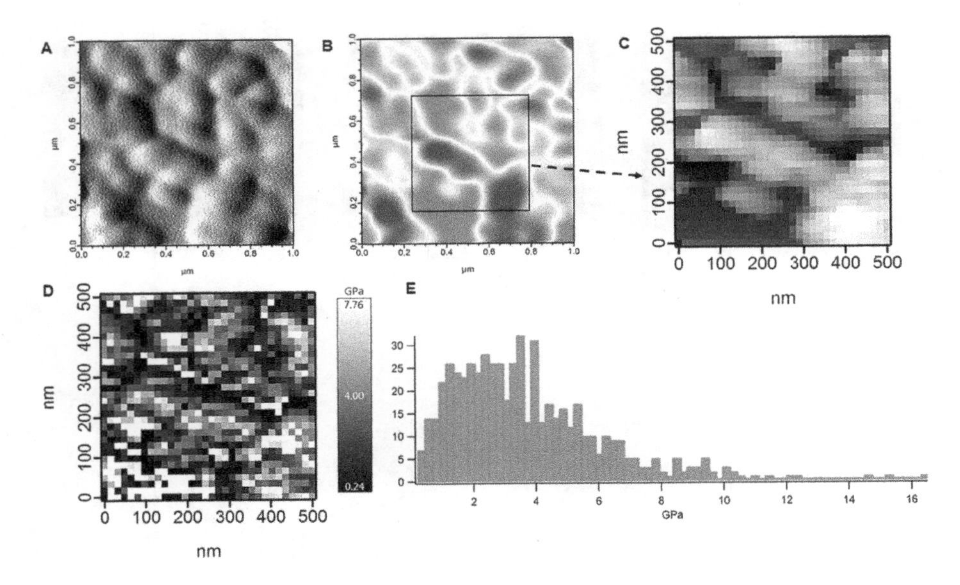

FIGURE 4.8 (a and b) AFM topography images along with the (c) height map and (d) stiffness map of Pd–Ag nanoparticles on the substrate after the stability test. (e) Histogram of the stiffness of Pd–Ag nanoparticles. The stiffness of most nanoparticles in the selected area in Figure S6-C ranges from 1 to 6 GPa. Reproduced with permission from Ref.[176] Copyright 2020, American Chemical Society.

(Figure 4.10c). To improve the energy efficiency of the electrochemical NRR, a gas-phase electrochemical cell is used, consisting of an MEA and GDLs (Figure 4.11). The cathode catalyst is prepared by mixing cleaned Pd–Ag nanoparticles in 0.675 μL of isopropanol (IPA) and 0.675 μL of Nafion solution (5% wt.) and the resulting solution is sonicated for 1 h and then spray-coated on the cation exchange membrane

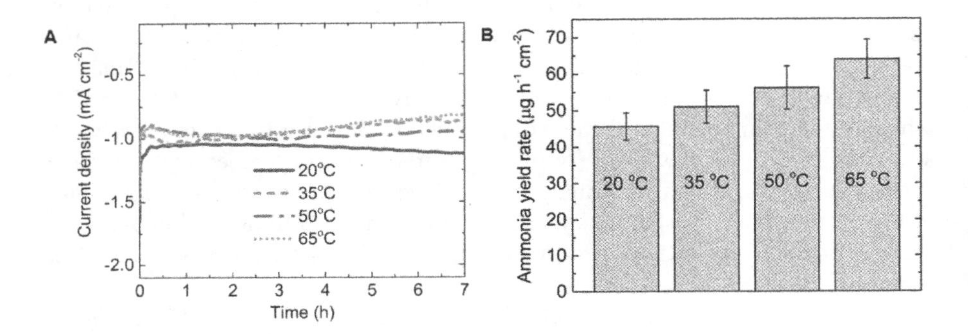

FIGURE 4.9 (a) CA tests of Pd–Ag nanoparticles at various temperatures at the applied potential of −0.6 V vs. RHE in N_2-saturated 0.5 M $LiClO_4$ (aq.) solution. (b) Ammonia yield rate using Pd–Ag nanoparticles at various temperatures at −0.6 V vs. RHE in 0.5 M $LiClO_4$ (aq.) solution. *(Continued)*

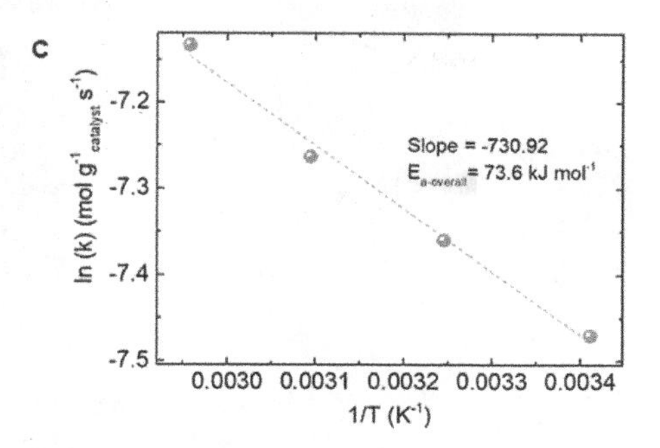

FIGURE 4.9 (Continued) (c) Arrhenius plot of the NRR rate ($\boldsymbol{k}$) using Pd–Ag nanocatalyst at various temperatures at the applied potential of −0.6 V. Reproduced with permission from Ref.[176] Copyright 2020, American Chemical Society.

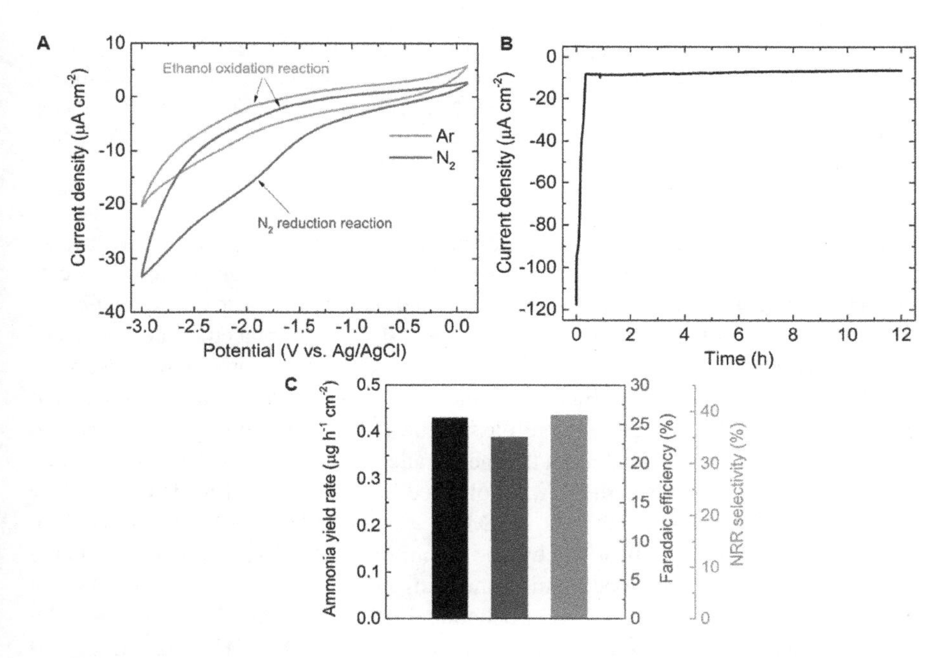

FIGURE 4.10 (a) CV curves of bimetallic Pd–Ag nanoparticles in Ar- and N_2-saturated 0.5 M $LiClO_4$ in THF solution at the scan rate of 10 mV s^{-1}. Ethanol (1% vol.) is used as a proton source. (b) CA result of Pd–Ag nanoparticles at the applied potential of −3.0 V vs. Ag/AgCl in N_2-saturated 0.5 M $LiClO_4$ in THF solution. Ethanol (1% vol.) is used as a proton source. (c) Ammonia yield rate, FE, and NRR selectivity of Pd–Ag nanoparticles at the applied potential of −3.0 V vs. Ag/AgCl in N_2-saturated 0.5 M $LiClO_4$ in THF solution. Ethanol (1% vol.) is used as a proton source. Selectivity is defined as the ratio of $(I_{N2} - I_{Ar})/I_{N2} \times 100$ at the applied potential of −3.0 V vs. Ag/AgCl. Reproduced with permission from Ref.[176] Copyright 2020, American Chemical Society.

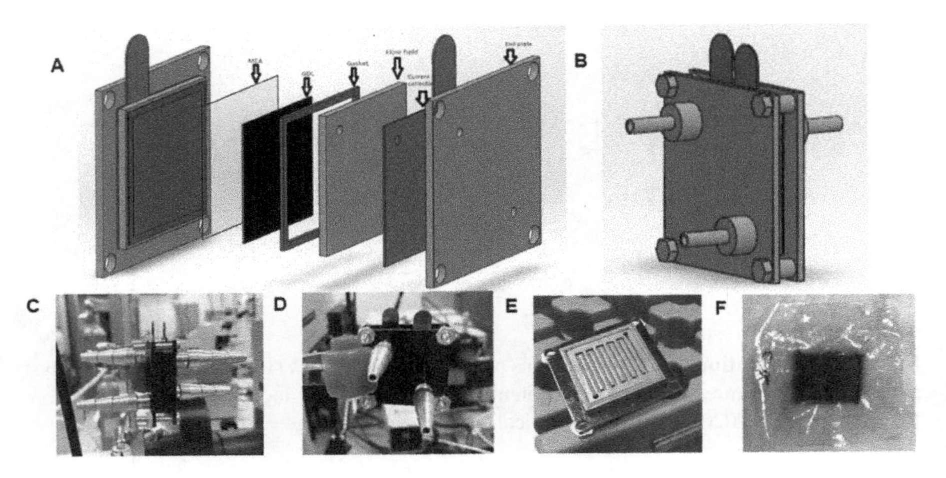

FIGURE 4.11 Schematic of the gas-phase setup for electrochemical NRR. (a) Various components of gas-phase setup, including MEA, GDL, gasket, flow field, current collector, and endplates. (b–d) Assembled gas-phase setup with one inlet and one outlet for each side (i.e., cathode and anode) for purging N_2 and H_2 gas. The outlet of the cathode side is connected to the absorber (0.001 M H_2SO_4) where the ammonia is trapped in the acid absorber for further quantification. (e) Serpentine flow field, and (f) MEA, catalyst-coated membrane is prepared via spray coating. Reproduced with permission from Ref.[176] Copyright 2020, American Chemical Society.

(Nafion-211). Pt nanoparticles were synthesized via the galvanic replacement process from the Ag nanocubes (AgNCs) template.[16] Pt nanoparticles were mixed with carbon black, sonicated for 1 h and then spray-coated on the other side of the membrane with a Pt loading of 0.2 mg cm^{-2}. Carbon paper GDLs were placed on both sides of the MEA with an effective area of 2.25 cm^2. The anode and the cathode were fed with pure H_2 and N_2, respectively. Electrochemical NRR was conducted at ambient lab temperature (20 °C), and gases were passed through a humidifier at 60 °C before entering the electrolyzer to maintain the membrane's humidity during measurements. Ammonia produced on the cathode was collected in an absorber (1 mM H_2SO_4, pH = 3). After electrochemical NRR tests, the MEA was soaked in 1 mM H_2SO_4 for 24 h to extract the ammonia absorbed by the membrane. By decreasing the distance between the cathode and anode and eliminating the liquid electrolyte, MEAs can reduce the ohmic losses in the electrochemical cell.[180–182] Pure H_2 is fed into the anode, where the hydrogen oxidation reaction (HOR) ($H_2 (g) \rightarrow 2H^+ + 2e^-$, $E^0_{RHE} = 0$ V) occurs on the Pt-based nanoparticles (Figure 4.12a).[176] HER is a competing reaction with the NRR at the cathode; H_2 gas from this reaction can be recovered and fed to the anode. The HOR overpotential is lower than that of OER on the Pt-based catalysts.[137] CA tests are conducted at lower applied potentials in the gas-phase than those of the liquid-phase system (Figure 4.12b). The same catalytic activity trend as liquid phase is observed in the gas phase, but with a lower ammonia yield rate (19.4 ± 2.1 μg cm^{-2} h^{-1}) and FE (7.9%) at an applied potential of −0.07 V with a comparable current density (1.15 mA cm^{-2} @−0.07 V in the gas-phase vs. 1.1 mA cm^{-2} @−0.6 V

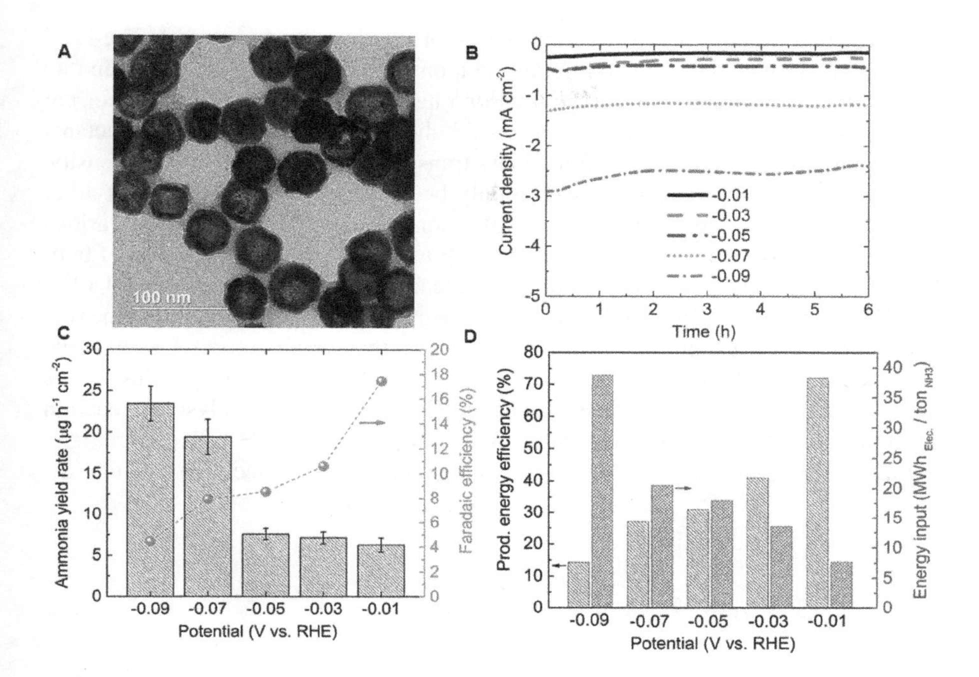

FIGURE 4.12 (a) Pt nanoparticles prepared via galvanic replacement technique from Ag nanoparticles used as an anode catalyst in electrochemical NRR in the gas-phase system. (b) CA results of Pd–Ag nanoparticles at a series of applied potentials in the gas-phase setup. (c) Ammonia yield rate and FE at various applied potentials using Pd–Ag nanoparticles in the gas-phase setup (MEA configuration). (d) Production energy efficiency and energy input at various applied potentials using Pd–Ag nanoparticles in the gas-phase system. Reproduced with permission from Ref.[176] Copyright 2020, American Chemical Society.

in the liquid phase) (Figure 4.12c). This performance corresponds to the energy efficiency of 27.1% and the energy input of 20.5 MWh per ton of ammonia, which is more efficient than the liquid phase due to lower ohmic resistances (Figure 4.12d). The highest energy efficiency achieved is 72.0% at an applied potential of −0.01 V, equivalent to an energy input of 7.7 MWh per ton of ammonia. This performance is achieved at a low current density of ~0.17± 0.03 mA cm^{-2}, comparable with the state-of-the-art Haber–Bosch process, which uses 7.8 MWh ton of ammonia (based on the natural gas as an input feedstock).[36]

To fully understand the role of mass transport resistances at the electrode surface in the H-cell, rotating disk electrode (RDE) measurements are conducted with nanoparticles deposited on a glassy carbon electrode.[183] The overall current density on the RDE is related to the kinetic-limited (i_k) and diffusion-limited current density (i_d). The RDE data are analyzed using the Koutecky–Levich (K-L) equation which is given by:

$$\frac{1}{i} = \frac{1}{i_k} + \frac{1}{i_d} \tag{4.3}$$

where i_d is proportional to the square root of rotation rate ($\omega^{\frac{1}{2}}$, rad/s). The LSV measurements are conducted at different rotation rates (0–3000 rpm) at the scan rate of 10 mV s^{-1} (Figure 4.13a). As the rotation rate increases, the value of the current density at −0.6 V vs. RHE reaches a plateau, indicating the transport of the reactants is no longer the limiting step (infinite mass transport) in the overall current density, and the rate of the half-reaction would only be limited by the slow kinetics at an electrode surface (Figure 4.13b). This is also consistent with the CA data at various rotation rates, where at the high rotation rate (i.e., 2500 rpm), the formation of bubbles at the electrode surface is limited due to the facile mass transport (Figure 4.14a). The K-L plot reveals a linear relationship between $1/i$ and $\omega^{-1/2}$, highlighting the role of kinetic- and diffusion-limited current density on the overall current density at the electrode surface (Figure 4.13c). To understand the role of mass transport loss in the H-cell system, we performed electrochemical NRR with similar catalyst and reaction conditions (e.g., catalyst loadings, applied potential, etc.) in the H-cell system under vigorous stirring and in the RDE system without stirring at an electrode rotation rate

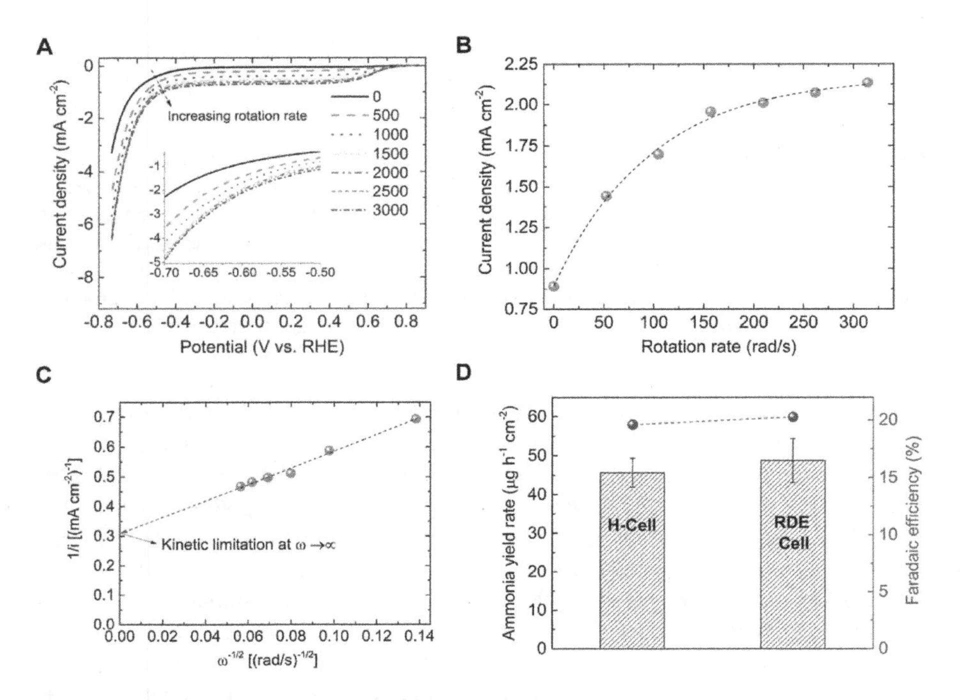

FIGURE 4.13 (a) LSV tests of Pd–Ag nanoparticles in an N_2-saturated 0.5 M $LiClO_4$ (aq.) solution in an RDE setup at various rotation rates (rpm) with a scan rate of 10 mV s^{-1}. (b) Current density at a potential of −0.6 V vs. RHE for various rotation rates. At higher rotation rates, a marginal increase in the current density is observed. (c) Koutecky–Levich plot from electrochemical NRR data on Pd–Ag nanocatalysts. (d) Ammonia yield rate and FE in an H-cell and RDE cell (2500 rpm) at −0.6 V vs. RHE in 0.5 M $LiClO_4$ (aq.) solution. Reproduced with permission from Ref.[176] Copyright 2020, American Chemical Society.

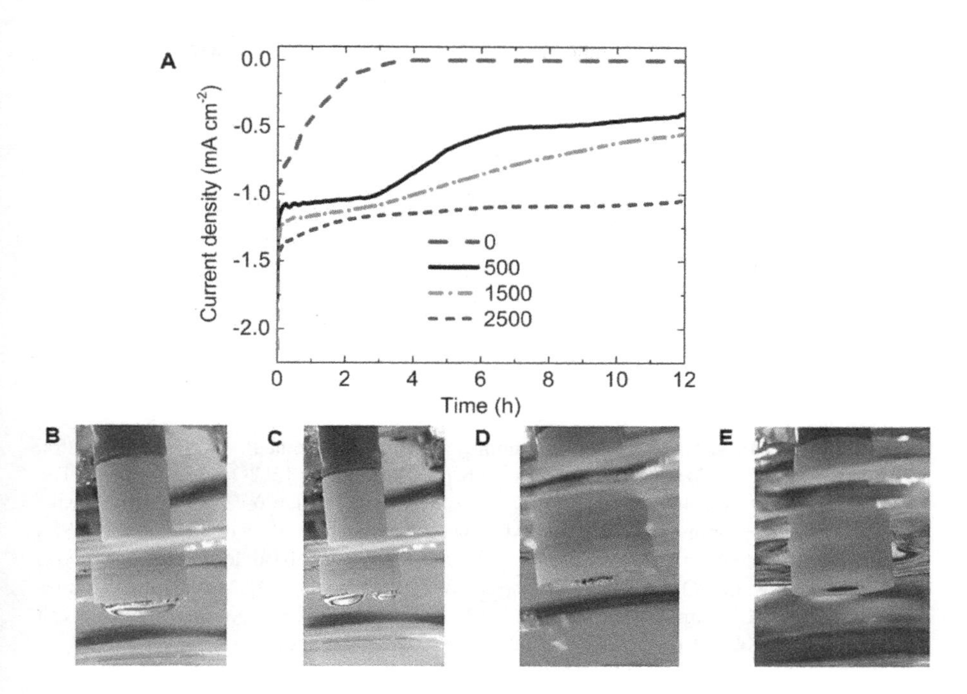

FIGURE 4.14 (a) CA tests at −0.6 V vs. RHE in 0.5 M $LiClO_4$ (aq.) solution at various rotation rates using Pd–Ag nanoparticles. (b–e) Images of the working electrode after the electrochemical NRR test at 0, 500, 1500, and 2500 rpm. At high rotation rates, the reactant mass transport loss is limited, as is evident from the lower amounts of bubbles with smaller sizes formed at the tip of the electrode. Reproduced with permission from Ref.[176] Copyright 2020, American Chemical Society.

of 2500 rpm. Comparable electrocatalytic NRR activity is observed in both set-ups, indicating that mass transport resistance in H-cell will be sufficiently suppressed if the reactor is intensely stirred under reaction conditions (Figure 4.14b, c, d, e).

Control experiments are carried out in which Argon (Ar) and isotopically labeled N_2 are used as feed gases.[114,145–147,184] $^{15}N_2$ (98 atom % ^{15}N) was purified before use in experiments by passing through an absorber (1 mM H_2SO_4) followed by deionized water to remove any NO_x and NH_3 contamination. The amounts of ammonia are measured with Nessler's method and ^1H NMR. The calibration curves for ^1H NMR quantification are provided in our prior studies.[114,184] No ammonia is produced in experiments with similar operating conditions (e.g., catalyst type, loading) to those in NRR experiments but with Ar gas and N_2 gas at the open-circuit voltage (OCV) (Figure 4.15a, b). Further, the doublet and triplet couplings of $^{15}N_2$ and $^{14}N_2$ with J-coupling constants of 72.9 Hz and 52.2 Hz obtained from ^1H NMR measurement and the quantitative agreement between the amounts of ammonia produced using $^{14}N_2$ and $^{15}N_2$ confirm that the supplied N_2 is the major source of ammonia formation in the system (Figure 4.15a). The amounts of ammonia measured by ^1H NMR in ^{15}NRR and ^{14}NRR experiments are 706.7 μM for $^{15}N_2$ and 769.3 μM for $^{14}N_2$.

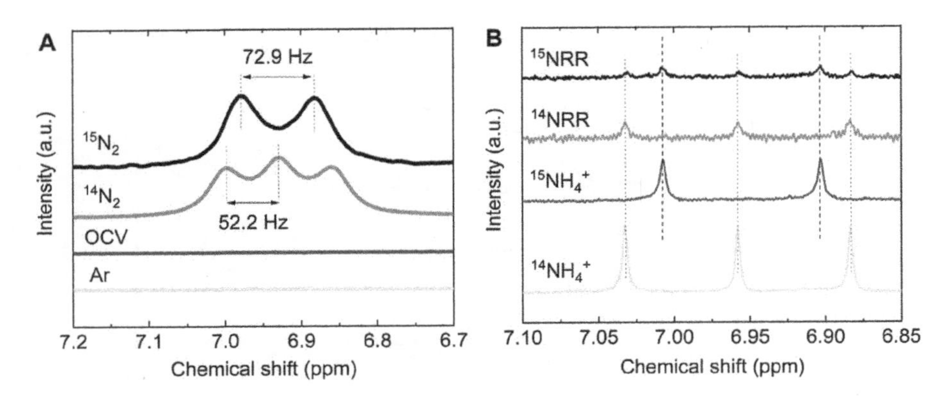

FIGURE 4.15 (a) ^1H NMR spectra of samples after electrochemical $^{15}N_2$ ($^{14}N_2$) reduction reaction at −0.6 V vs. RHE using Pd–Ag nanoparticles in 0.5 M $LiClO_4$ (aq.) solution. The control experiments spectra (OCV and Ar) are also included, which reveal no distinct peaks associated with the ammonium. (b) ^1H NMR spectra of samples after electrochemical $^{15}N_2$ ($^{14}N_2$) reduction reaction at −3.0 V vs. Ag/AgCl using Pd–Ag nanoparticles in the nonaqueous electrolyte (0.5 M $LiClO_4$ in THF and 1% ethanol). Standard $^{15}NH_4^+$ and $^{14}NH_4^+$ samples (25 μM) were used for ammonia quantification in the nonaqueous solution. Reproduced with permission from Ref.[176] Copyright 2020, American Chemical Society.

These values are in close agreement with the amounts of ammonia measured by Nessler's method (804.9 ± 65.3 μM). The amounts of ammonia reported in this work are remarkably higher than the amount of ammonia that can be produced by even maximum contamination levels of various impurities (e.g., NO_x) that might be present in ultra-high-purity (UHP) $^{14}N_2$ used for experiments. This also highlights the importance of operating N_2 electrolysis at high current densities to achieve high ammonia yield rates to rule out the false-positive results of ammonia formation by gas impurities.

The amounts of ammonia produced in the nonaqueous electrolyte were also comparable in the ^{14}NRR and ^{15}NRR (1.52 ± 0.37 μM) (Figure 4.15b). Some $^{14}NH_4^+$ is observed in the ^{15}NRR experiment, which is attributed to various sources of contamination in measurements.[145]

This section highlighted the importance of cell design and optimization in addition to the design of selective catalysts for high-performance N_2 electrolysis for ammonia synthesis.

5 Plasma-Enabled Nitrogen Fixation

5.1 INTRODUCTION

Plasma is an ionized gas. A well-known example of a plasma state is the Sun, which is revealed in the form of lightning. Plasma can be created by supplying energy to a gas. There is a difference between fusion plasmas and gas discharge plasmas. Fusion plasmas operate at extremely high temperatures (millions of degrees) to mimic the Sun's conditions to realize nuclear fusion as a future energy source. Gas discharge plasmas, on the other hand, operate at a much lower temperature, close to room temperature, and are created by applying electrical energy to a gas. The latter is potentially interesting for use in renewable energy conversion applications. Plasma allows the activation of small molecules, where the gas is not heated, and the applied electrical energy will selectively heat the electrons due to their small mass. These energetic electrons will collide with the gas molecules (e.g., N_2), causing excitation, ionization, and dissociation. The excited species, ions, and radicals will react further, generating new molecules. There is thermal nonequilibrium between the highly energetic electrons (i.e., tens of thousands K) and the gas molecules (e.g., at room temperature up to a maximum of a few thousand K). This allows thermodynamically unfavorable or energy-intensive chemical reactions, such as N_2 fixation, to proceed in an energy-efficient way using renewable electricity. In thermal plasmas, all species (electrons, gas molecules, excited species, ions) have the same temperature, but as expected, they are less energy-efficient for small molecule activations.[192,193]

Plasma-catalytic nitrogen fixation is the redox reactions of plasma-generated ionized N_2 with H_2 or O_2 through three steps: 1) the collision of N_2 with plasma-generated highly energetic electrons to form vibrationally excited nitrogen species (N_2^*), 2) the activation of molecular O_2 or H_2 over catalysts, and 3) reactions between N_2^* and adsorbed O_2 or H_2 species to form ammonia or nitrogen oxides.[194,195] In non-thermal plasma ammonia synthesis, various plasma sources (e.g., dielectric barrier discharge (DBD)) are used to generate plasma at low temperature and pressure.[196] DBD has received more attention than other plasma sources as it enables the control of the average energy of discharged electrons by adjusting the discharge gas width and gas pressure.[194] Early studies used a nozzle-type plasma jet to produce ammonium ions in water.[197] Recent studies further revealed that UV irradiation could provide additional hydrogen donors to the plasma-assisted ammonia synthesis process by exciting water molecules adsorbed on the catalyst surface.[198–200] In addition, ammonia can be synthesized in plasma chambers even without the use of a metal catalyst.[100,201] In plasma-enabled ammonia synthesis, nitrogen hydrogenation pathways occur on a catalyst's surface through a dissociative mechanism similar to the thermocatalytic ammonia synthesis.

5.2 DIFFERENT PLASMA TYPES FOR ENERGY APPLICATIONS

A gas discharge plasma is generated by applying an electric potential difference between two electrodes placed in a gas. The gas pressure can be varied from a few Torr up to >1 atm. The potential difference can be direct current (DC), alternating current (AC), ranging from 50 Hz over kHz to MHz (radio frequency, RF), or pulsed. In addition, the electrical energy can be supplied in various ways, such as inductively coupled plasma (ICP) or microwaves (MWs). There are three types of plasmas most often studied for small molecule activation, namely dielectric barrier discharges (DBDs), MW plasmas, and gliding arc (GA) discharges. It is noted that other plasma types are explored for these applications as well, such as nanosecond (ns) pulsed discharges, spark discharges, corona discharges, and atmospheric pressure glow discharges.[202]

DBD operates at atmospheric pressure and is generated by applying an AC potential difference between two electrodes, at least one of two electrodes covered by a dielectric barrier. This limits the amount of charge transported between two electrodes, and the electric current, preventing the discharge from undergoing a transition into a thermal regime. The most studied electrodes in this type of plasma are two concentric cylindrical electrodes (Figure 5.1a). The reactor is comprised of an inner electrode surrounded by a dielectric tube with a gap in the size of a few millimeters. The second electrode is typically a mesh or foil that is wrapped around the dielectric tube. One of the electrodes is connected to a power supply, while the other electrode is grounded. The gas flows in from one side and is gradually converted along its way through the gap between the inner electrode and dielectric tube, similar to the process in a plug flow reactor, and flows out from the other side. The industrial implementation of DBD

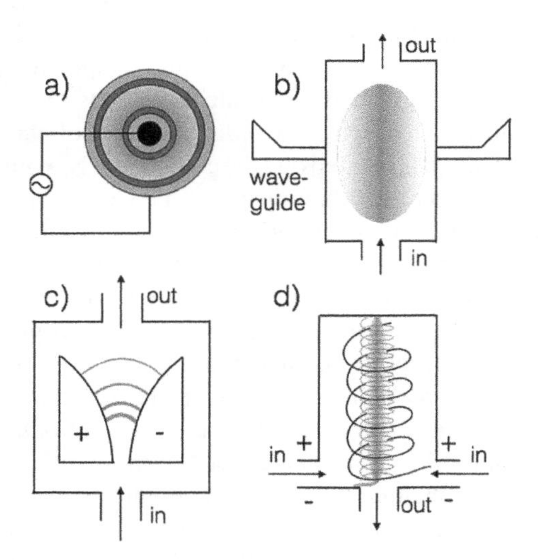

FIGURE 5.1 Schematic illustration of the three plasma reactors most often used for small molecule activation (e.g., N_2 fixation) applications (a) DBD, (b) MW plasma, and GA discharge, in (c) classical configuration and (d) cylindrical geometry, called GAP. Reprinted with permission from Ref.[192] Copyright 2018 American Chemical Society.

is realized by placing a large number of DBD reactors in parallel, which operates at atmospheric pressure. The energy efficiency for small molecule activation (e.g., N_2 fixation) is typically very low. The production conversion efficiency could be improved by incorporating a catalytic material to enable the selective production of targeted molecules in the process called "plasma catalysis." Plasma provides a very reactive environment due to various chemical species (electrons, molecules, atoms, radicals, ions, and excited species), but it is not selective in the production of targeted molecules. Plasma catalysis combines the high reactivity of plasma with a catalyst material's selectivity so that targeted species can be formed with high product yield and selectivity. However, more research is needed to design catalysts tailored to the plasma environment effectively.[192]

A microwave (MW) plasma is generated by applying MWs, which is electromagnetic radiation with a frequency between 300 MHz and 10 GHz, to a gas, without using electrodes. There are different types of MW plasmas, including cavity-induced plasmas, free expanding atmospheric plasma torches, electron cyclotron resonance plasmas, and surface wave discharges. The latter type is most frequently used for small molecule applications. The gas flows through a quartz tube, transparent to MW radiation, intersecting with a rectangular waveguide, to initiate the discharge (Figure 5.1b). The MWs penetrate along the interface between the quartz tube and the plasma column, and the plasma absorbs the wave energy. MW plasmas can operate at pressures as low as ten mTorr up to atmospheric pressure. When the system operates at low pressure, it achieves a relatively high energy efficiency. Still, when the system operates above 0.1 atm, they approach thermal equilibrium, with a few thousand K gas temperatures, thereby remarkably reducing the system's energy efficiency.[192]

A Gliding Arc (GA) discharge is a transient type of arc discharge. A classical GA discharge is formed between two flat diverging electrodes (Figure 5.1c). The arc is initiated at the shortest interelectrode distance. Under the influence of the gas blast, which flows along the electrodes, the arc "glides" toward a larger interelectrode distance until it extinguishes. A new arc is created at the shortest interelectrode distance.[192] This type of (two-dimensional) GA discharge yields only limited product formation because a large fraction of the gas does not pass through the arc discharge. Therefore, other types of (three-dimensional) GA discharges have been designed, such as a gliding arc plasmatron (GAP) and a rotating GA, operating between cylindrical electrodes (Figure 5.1d). The cylindrical reactor body operates as a cathode (powered electrode), while the reactor outlet acts as an anode and is grounded. The gas enters tangentially between the two cylindrical electrodes. When the outlet diameter is considerably smaller than the reactor body's diameter, the gas flows in an outer vortex toward the reactor body's upper part. Subsequently, it will flow back in a reverse inner vortex with a smaller diameter because it loses momentum, and thus, it can leave the reactor through the outlet. The arc is again initiated at the shortest interelectrode distance. It expands until the upper part of the reactor, rotating around the reactor's axis, stabilizes around the center after about 1 ms. In the ideal scenario, the inner gas vortex passes through this stabilized arc, allowing a larger fraction of the gas to be converted than in a classical two-dimensional GA discharge. Nevertheless, the fraction of gas passing through the active arc is still limited.[192]

The GA discharge operates at atmospheric pressure, making it also suitable for industrial implementation. Furthermore, it exhibits relatively high energy efficiency for small molecule activation. However, the gas temperature is also relatively high (i.e., a few thousand K), limiting the energy efficiency, similar to a MW plasma, since the plasma approaches thermal equilibrium.

5.3 PLASMA-BASED N_2 FIXATION

The theoretical limit of plasma-based N_2 fixation's energy consumption is more than 2.5 times lower than that of the Haber–Bosch process. Plasma-based N_2 fixation can be conducted at atmospheric temperature and pressure, which contrasts to the high pressure and temperature requirements of the Haber–Bosch process. Thus, Plasma-based N_2 fixation could result in achieving lower operational costs and more increased process safety.[192] Two processes can be studied in plasma-based N_2 fixation: 1) NH_3 synthesis from N_2 and H_2, and 2) NO_x formation from the air ($N_2 + O_2$). Ammonia synthesis has been demonstrated in a single-stage plasma catalysis reactor packed with Ru-Mg/γ-Al_2O_3 pellets at different temperatures.[203] It was shown that the optimum ratio in plasma catalysis is the N_2:H_4 ratio of 4:1. At temperatures higher than 200 °C when the catalyst reaches the light-off temperature, the plasma's effect becomes dominant (810 ppm NH_3 compared to 10 ppm without plasma; see Figure 5.2).[203] The energy yield obtained in these experiments was 25–30 g of NH_3/kWh.[203] The production rate of ammonia has been examined using various transition metal catalysts, including Fe, Ru, Co, Ni, Pt.[204] In these experiments, 100 mg of material was packed into the plasma-catalytic reactor, and the total flow rate of reactants was varied between 10 and 50 mL/min. The experiments were conducted at 438 K using a feed composition of $N_2/H_2 = 2$ and a plasma power of 10 W. It was demonstrated that among various transition metal catalysts, Ni has the highest production rates as a function of the residence time of reactants in the DBD reactor. However, for all cases, production rates decrease as the residence time of reactants (W/F) increases (Figure 5.3).[204]

Plasma electrolytic system can be demonstrated in the catalyst-free environment.[100,198] In this system, a plasma is ignited in the gas gap between a stainless-steel capillary tube and electrolyte solution (H_2SO_4 (aq.)) surface (Figure 5.4a), and ammonia is synthesized by the solvated electrons produced at the plasma-water interface. In this catalyst-free system, the ammonia selectivity reached 100% at low currents (1 and 2 mA) while decreasing at higher current densities (e.g., 4, 6, 8 mA) (Figure 5.4b). The highest ammonia formation rate of 0.44 mg h^{-1} (7.2×10^{-7} mol cm^{-2} s^{-1}) was observed at 2 mA applied current. Stability is still a significant issue as the ammonia selectivity decreased from 60% at 5 min to 30% after 45 min experiments, and then to 25% after 5 h of electrolysis at 6 mA applied current (Figure 5.4c). This plasma electrolytic system enables the production of ammonia with a very high selectivity and formation rate. In addition, the large energy consumption of the plasma system (~500 V was required to generate and sustain the plasma) is a significant hurdle that may prevent its large-scale adoption. The system has a power consumption of 2,270 kWh per kg of NH_3, which corresponds to an energy conversion

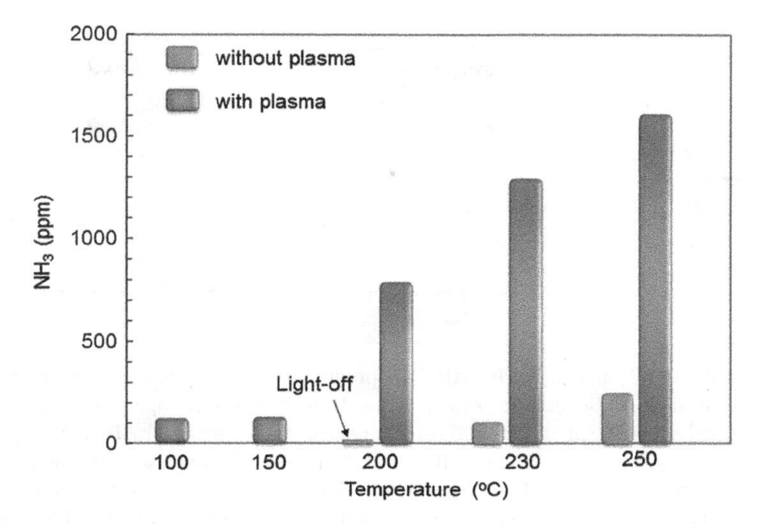

FIGURE 5.2 NH_3 concentration obtained in a packed-bed DBD plasma reactor with Ru – Mg/γ-Al_2O_3 packing at a gas flow rate of 2 L/min and an H_2:N_2 ratio of 1:4, compared with and without plasma at different temperatures. Reprinted with permission from Ref.[203]

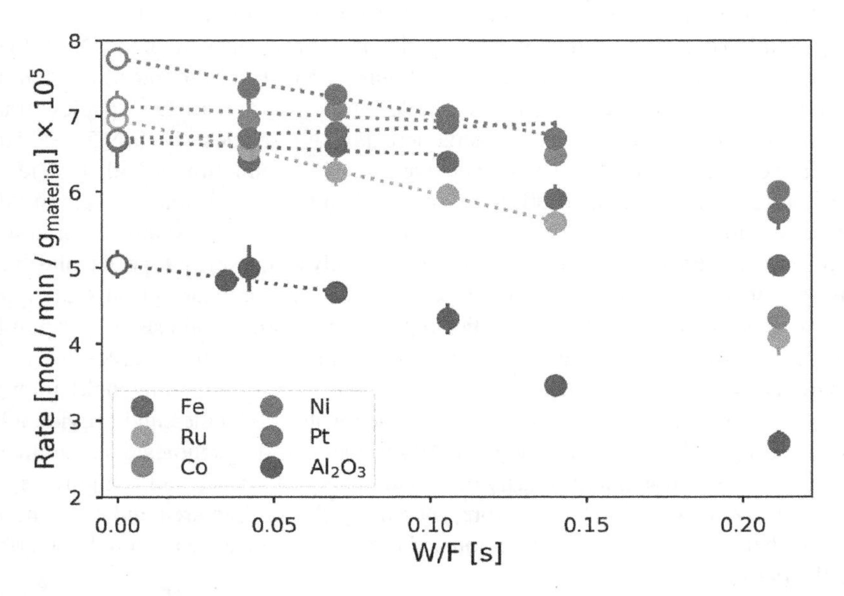

FIGURE 5.3 Ammonia production rates as a function of the residence time of reactants (W/F) in the DBD reactor. Initial rates (plotted as open circles) were extracted by extrapolating the observed rates to W/F = 0 using uncertainty weighted linear regression (shown by dotted lines). Reaction conditions: 438 K, 10 W, inlet N_2:H_2 = 2:1. Error bars indicate the standard deviation of the rates. Reprinted with permission from Ref.[204]

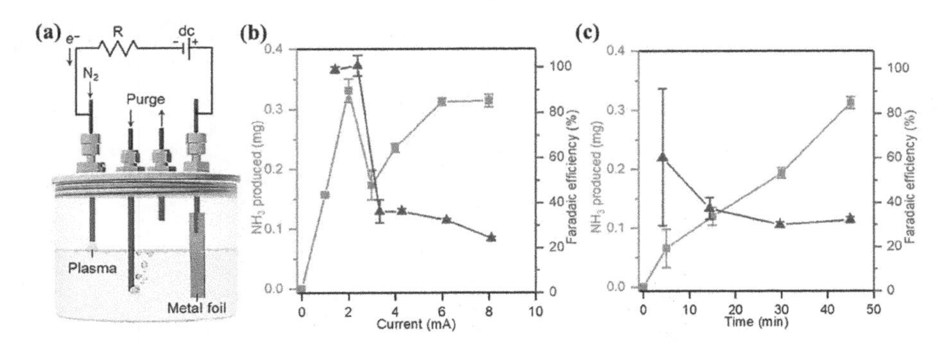

FIGURE 5.4 Catalyst-free, electrolytic NH_3 production from N_2 and water using a plasma electrolytic system. (a) Schematic of the plasma electrolytic system operated by a DC power supply and galvanostatically controlled using a resistor (R) in series. (b) Total NH_3 produced and corresponding Faradaic efficiency after different processing times at 6 mA and pH 3.5. No NH_3 is produced at 0 min based on the untreated electrolyte solution. (c) Total NH_3 produced and corresponding Faradaic efficiency as a function of current after 45 min electrolysis at pH 3.5. No NH_3 is produced at 0 mA based on the control experiment. Reprinted with permission from Ref.[204]

efficiency of approximately 0.25%. This is compared with the conventional Haber–Bosch process using natural gas, coal, or fuel oil as feedstocks requires only 9–13 kWh per kg of NH_3 with an energy conversion efficiency of 60%.[99]

N_2 fixation can occur in the oxidation pathway to produce nitric oxide (NO). In the recently reported solid-state electrolyzer (Figure 5.5a), H_2O is reduced to H_2 in the cathode while NO is produced in the anode by the reaction of O^{2-} species (transported via the electrolyte) and plasma-activated N_2 species (Figure 5.5b). High Faradaic efficiencies up to 93% are achieved for NO production at 650 °C, and NO concentration is more than 1000 times greater than the equilibrium concentration at the same temperature and pressure.[205] In another study, the co-synthesis of ammonium, nitrite, and nitrate was studied using an advanced spray-type jet plasma.[200] This was accomplished by the interactions between the gas-phase plasma and liquid surface dissociation. Two recent studies reported ammonia synthesis by purging the nitrogen plasma gas into water under UV light irradiation, which successfully promoted the ammonium synthesis rate.[198,199,206] The UV irradiation could improve ammonia synthesis via excitation of the surface water molecules and provide additional hydrogen donors to the reactions. Additionally, findings indicate that ammonia synthesis is a reaction that primarily occurs at the liquid surface and is highly dependent on the reaction area.[198] Therefore, increasing the surface area and reducing the distance between the plasma generation and the water surface are essential to improving the performance.

An industrial process for plasma-based NO_x synthesis might be more appealing than plasma-based ammonia synthesis as air can be utilized directly as the feed gas. The primary challenge with the plasma-based N_2 fixation is the low energy conversion efficiency of the process. However, plasma-based N_2 fixation can potentially adopt for distributed production plants by integrating with renewable energy sources.

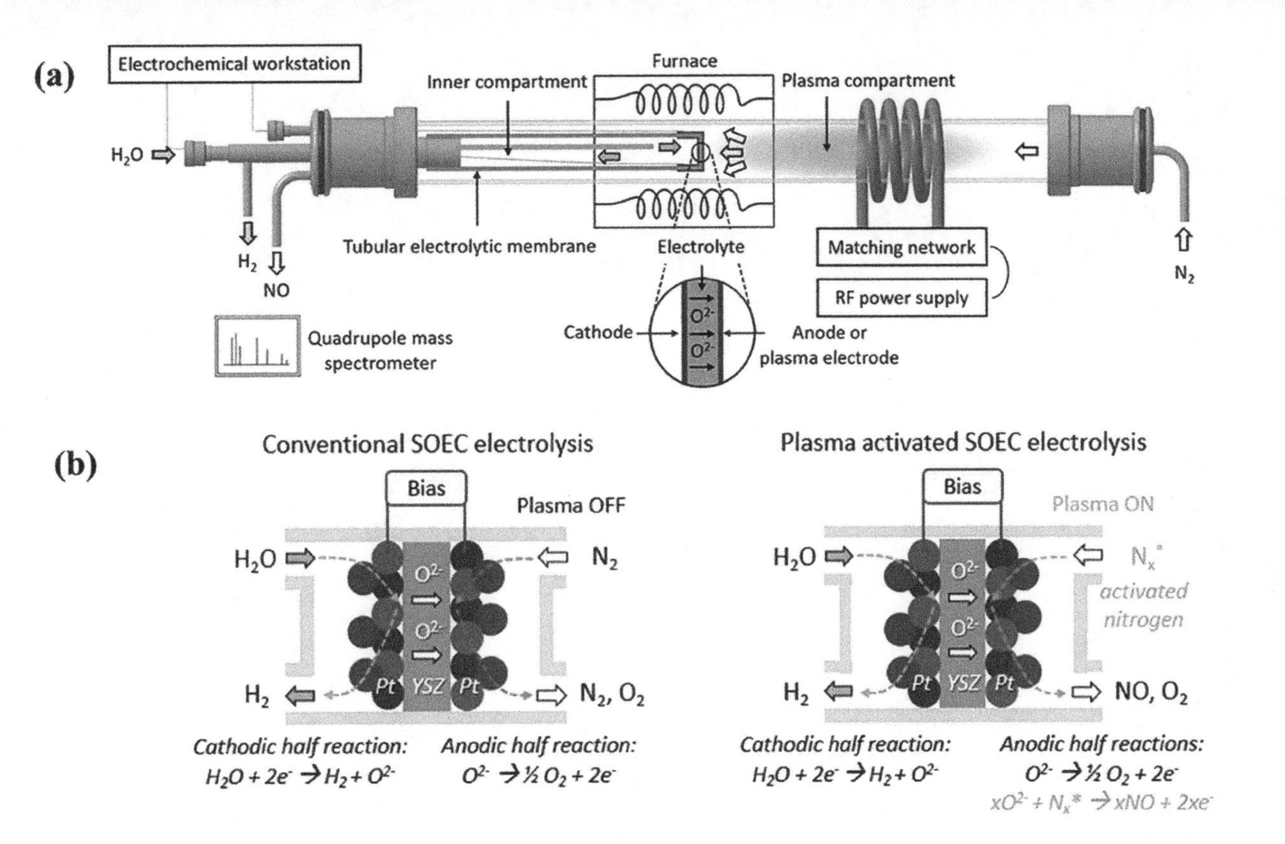

FIGURE 5.5 (a) Schematic representation of hybrid plasma-activated solid oxide electrolyte cell (SOEC) reactor setup. (b) Operation principles of conventional (left) and activated (right) SOEC electrolysis. In this system, H_2O is reduced to H_2 in the cathode of a solid oxide electrolyzer while NO is produced in the anode by the reaction of O^{2-} species (transported via the electrolyte) and plasma-activated N_2 species. Reprinted with permission from Ref.[205] Copyright 2019 American Chemical Society.

Plasma can be switched on and off very quickly to follow the renewable energy supply as it requires no preheating or long stabilization times and no cool-down times.[192] The conversion reactions start immediately after plasma ignition. This makes plasma technology well suited for converting intermittent renewable energy into fuels or high-value chemicals. Furthermore, repeated on/off cycles do not damage the plasma reactors.

6 Photocatalytic Nitrogen Fixation

6.1 INTRODUCTION

Photocatalytic nitrogen fixation for "green ammonia" synthesis, using water as a reducing reagent and sunlight under ambient conditions provides an attractive route for clean, sustainable, and distributed ammonia synthesis. This approach might be an alternative to the Haber–Bosch process, particularly in remote areas where access to the centralized infrastructure or pipelines is limited.[207] An Indian soil scientist first reported photocatalytic N_2 fixation.[208] Schrauzer and Guth then investigated the performance of metal oxides (e.g., TiO_2) photocatalysts for nitrogen fixation with iron-doped titanium oxide was found to be an effective catalyst.[209] This informed more research to understand the role of Fe in promoting photocatalytic ammonia synthesis. Various other semiconductor materials were tested for photocatalytic ammonia synthesis, including transition metal oxides, metal sulfides, bismuth oxyhalides, carbonaceous materials, layered double hydroxides, biomimetic photocatalysts, and biohybrid complexes.[162,210–212] Prior studies for photocatalytic N_2 fixation using various photocatalysts are presented in Figure 6.1. The reported photocatalysts enabled the conversion of solar energy into high-value chemicals without the need for additional energy input or sacrificial agents. However, limited solar spectrum utilization, fast charge recombination, and sluggish surface reaction kinetics hinder photocatalytic performance.[213] Photocatalytic N_2 fixation to NH_3 using water as the reducing agent is incredibly challenging, being a six electron, six proton reduction process. Besides, achieving adequate coverages of adsorbed N_2 is a further bottleneck, limiting photocatalytic nitrogen reduction reaction (NRR) activity. Research on photochemical ammonia synthesis has primarily focused on catalysts design. Engineering titanium oxide, bismuth oxyhalides, and layered double hydroxides for enhanced visible-light-driven photocatalytic NRR activity have been reported in literature.[213–215] Morphology control, including plane, corner, edge sites, porous structures, etc. and crystal manipulation, including amorphous layers, lattice strain, etc. are common strategies used to create highly exposed unsaturated metal centers, thus promoting N_2 adsorption and electron transfer processes on the catalysts' surface (Figure 6.2).[161,184,216] The introduction of lattice vacancy (oxygen and metal vacancies) and heteroatom doping are effective approaches for promoting NRR activity. The introduction of metal complexes and single metal atoms sites helps lower key intermediates' formation energy in NH_3 synthesis. For instance, by incorporating oxygen vacancies in defect-abundant titanium oxide or layered double hydroxides, photocatalytic nitrogen fixation under solar irradiation at wavelengths up to 700 nm has been observed.[217–219] It is imperative to engineer photocatalysts containing multifunctional active sites to suppress charge recombination and undesirable side reactions (ammonia oxidation, nitrogen

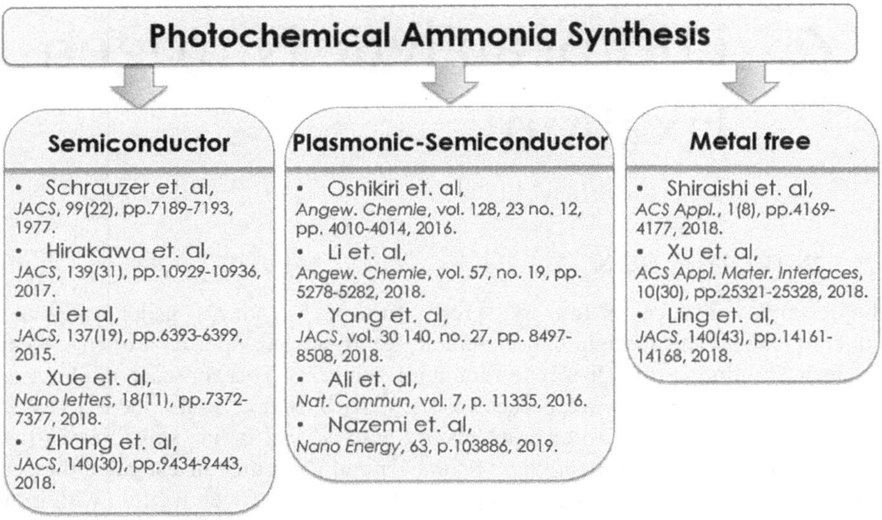

Photochemical Ammonia Synthesis

Semiconductor

- Schrauzer et. al, *JACS*, 99(22), pp.7189-7193, 1977.
- Hirakawa et. al, *JACS*, 139(31), pp.10929-10936, 2017.
- Li et al, *JACS*, 137(19), pp.6393-6399, 2015.
- Xue et. al, *Nano letters*, 18(11), pp.7372-7377, 2018.
- Zhang et. al, *JACS*, 140(30), pp.9434-9443, 2018.

Plasmonic-Semiconductor

- Oshikiri et. al, *Angew. Chemie*, vol. 128, 23 no. 12, pp. 4010-4014, 2016.
- Li et. al, *Angew. Chemie*, vol. 57, no. 19, pp. 5278-5282, 2018.
- Yang et. al, *JACS*, vol. 30 140, no. 27, pp. 8497-8508, 2018.
- Ali et. al, *Nat. Commun*, vol. 7, p. 11335, 2016.
- Nazemi et. al, *Nano Energy*, 63, p.103886, 2019.

Metal free

- Shiraishi et. al, *ACS Appl.*, 1(8), pp.4169-4177, 2018.
- Xu et. al, *ACS Appl. Mater. Interfaces*, 10(30), pp.25321-25328, 2018.
- Ling et. al, *JACS*, 140(43), pp.14161-14168, 2018.

FIGURE 6.1 Background of various photochemical approaches for nitrogen fixation in the literature. The photocatalyst design will be categorized as semiconductor, plasmonic-semiconductor, and metal-free photocatalysts. Selective studies under each category are provided.

oxidation, hydrogen evolution, etc.) during photocatalytic NRR. Photocatalytic N_2 oxidation from water and oxygen under ambient conditions might be an effective approach to produce nitric acid.[220,221] This area of research has received less attention than the nitrogen reduction pathway. Still, it could potentially be a promising pathway for nitrogen fixation as the competing hydrogen evolution reaction (HER) will not affect the photocatalytic nitrogen oxidation activity. Additionally, one of the products of this reaction is nitric oxide (NO), which is in the gas form under ambient conditions. Therefore, the separation of products in the aqueous solution is no longer a challenge. In addition to the catalyst engineering to enhance the surface catalytic reaction involved in N_2 fixation, system engineering is essential to achieve the efficient mass transfer of reactants (e.g., N_2 gas) to catalytic sites. Since gas molecules such as N_2, O_2, and CO_2 have poor solubility and diffusion coefficient in water, photocatalytic activity can be rate limited by diffusion processes in conventional aqueous phase reaction systems as shown in the right panel of Figure 6.2. To overcome these solubility and diffusion limitations, the design of gas-phase systems, where N_2 molecules can be directly supplied on the catalyst surface. In such systems, reactant gases (N_2 or air) are continuously introduced at the liquid–solid interface using gas diffusion layer (GDL). Since the concentration of nitrogen in the gas phase is about 140 times higher than the saturation concentration of dissolved nitrogen in the water, gas-phase systems comprised of GDL offer a much higher local nitrogen concentration on the surface of catalysts, thereby potentially resulting in much higher NH_3 production rates.

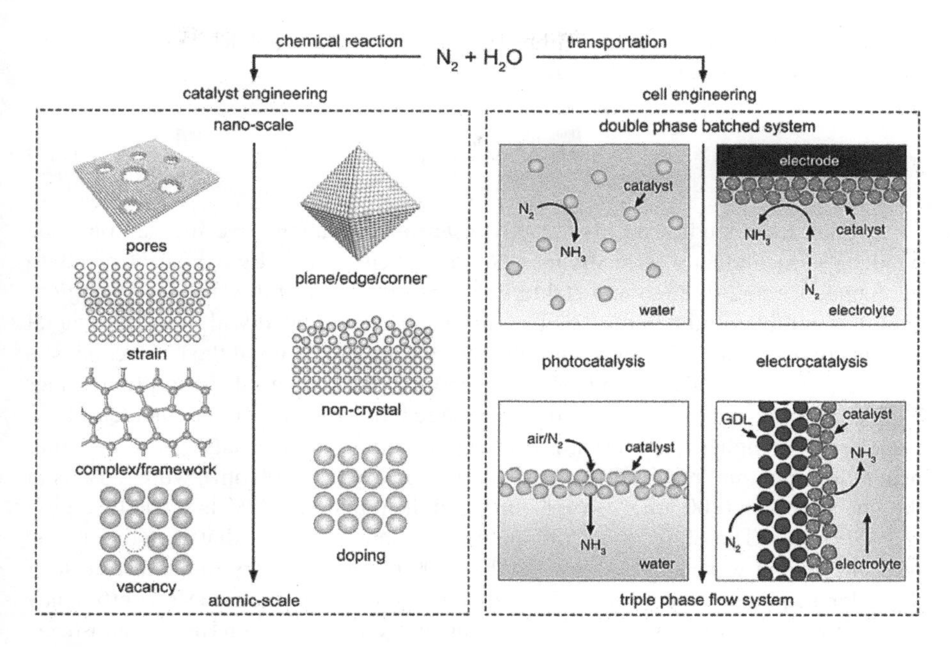

FIGURE 6.2 Schematic illustration of catalyst and cell engineering developed in photocatalytic and electrocatalytic NRR: (left) size comparison of various types of catalyst engineering, including pores, plane/edge/corner, vacancy, doping, etc. for enhanced surface chemical reaction and (right) cell engineering from liquid-phase (H-cell system) to gas-phase systems comprised of GDL to promote the interfacial nitrogen transportation. Reprinted with permission from Ref.[104]

6.2 DESIGN OF HYBRID PHOTOCATALYSTS

To date, almost all photocatalysts for NRR suffer from poor selectivity, low activity due to the difficulty in breaking the strong N≡N bond, competition with the more favorable HER, and inefficient utilization of the solar spectrum. Superior light absorption, efficient photo-excited charge separation, and transfer, and the ability to drive photocatalytic redox reactions are the key elements to the rational design of an efficient photocatalyst for the NRR.[222] The photocatalytic redox system using water as a reducing reagent comprises water oxidation at the valence band through the reaction of water with photo-generated holes (h^+) (Equation 6.1) and NRR at the conduction band through the reaction of N_2, protons (H^+), and photo-generated electrons (Equation 6.2):[209]

$$3H_2O + 6h^+ \rightarrow \frac{3}{2}O_2 + 6H^+ \tag{6.1}$$

$$N_2 + 6H^+ + 6e^- \rightarrow 2NH_3 \tag{6.2}$$

The overall reaction, with the Gibbs free energy (ΔG^o) of 339 kJ mol^{-1} for ammonia formation from water and N_2, is:

$$N_2 + 3H_2O \rightarrow 2NH_3 + \frac{3}{2}O_2 \tag{6.3}$$

To date, most studies on photocatalytic nitrogen fixation have focused on wide bandgap semiconductors (usually metal oxides) such as TiO_2 (Eg = 3.2 eV), owing to its abundance and its high and stable light absorption capability in the UV region. TiO_2 has exhibited moderate NRR activity when incorporated with transition metal dopants (e.g., Fe, Mo, Ru) or by introducing oxygen vacancies at the photocatalyst's surface.[209,219,223–226] Wide bandgap semiconductors show weak absorption under visible and IR light and, therefore, low NRR activity when the incident light wavelength is higher than 400 nm. It is important to note that only 5% of the solar irradiation spectrum received on earth's surface is UV ($\lambda < 400$ nm), while 50% is in the visible region (400 nm $< \lambda <$ 800 nm), and the remaining 45% is in the IR region ($\lambda >$ 800 nm). Therefore, the development of a photocatalyst that is active in the incident photon wavelength of greater than 400 nm is necessary to harvest most of the solar irradiation spectrum and enhance the solar-to-ammonia (STA) efficiency (η_{STA}). It was recently demonstrated that using oxygen vacancy rich BiOBr nanosheets can enhance the photocatalytic NRR activity under visible light in water. Yet, there is room to improve the apparent quantum efficiency (η_{AQE}).[227–229] Under the stimulus of light, some noble metal nanoparticles, such as Au and Ag, undergo a process called localized surface plasmon resonance (LSPR), which is the collective oscillation of free electrons at the material's surface.[222] The strong electromagnetic plasmon fields generated under light illumination can be utilized for various applications in photocatalysis and sensing.[113,160,230–234] The placement of plasmonic nanostructures in proximity (<10 nm) to each other results in an enhanced electromagnetic field in the interparticle gap.[235] Hybrid Au-Ag_2O nanocages are synthesized through a facile oxidation process, allowing for the coupling of plasmonic enhancement of Au nanoparticles with visible-light active p-type Ag_2O to achieve superior photocatalytic NRR performance under ambient conditions using water as an electron donor and sunlight without using sacrificial reagents. While several studies on photocatalytic NRR using hybrid plasmonic-semiconductor photocatalysts have been perfor med,[117,161,163,236] very few have coupled a plasmonic metal and semiconductor that are both active under visible light.[160] Combining the two increases the photo-generated electrons' concentration upon visible light illumination for photocatalytic NRR. In a photochemical system, atmospheric N_2 is converted to NH_3 under ambient conditions using visible light, and water as an electron donor. As the work function of the p-type semiconductor (Ag_2O) is higher than that of the Au plasmonic metal ($\varnothing_S > \varnothing_M$), a Schottky barrier is formed where the photo-generated holes of Ag_2O are transferred to the Au plasmonic metal at the metal–semiconductor interface so that the Fermi levels in the metal and semiconductor are aligned.[237,238] Energy bands in Ag_2O bend downward to match the chemical potentials. Photo-generated electron transfer from the conduction band of Ag_2O to the surface of the Au nanoparticles occurs, or so-called "hot electrons" are injected from the Au nanoparticle surface to the conduction band of Ag_2O. Both electron transfer processes are possible and competitive. It has been demonstrated that transition metal loaded titania results in smaller ammonia yield than bare titania if Ti^{3+} active sites on the surface are blocked by transition metals, giving experimental proof of these competing phenomenon.[219] In our nanocages, Au and Ag_2O nanoparticles are settled adjacent to each other, whereby both can act as an active site in the interior or at the exterior surface of the nanocages for nitrogen adsorption and reduction, enabling us to achieve superior selectivity and efficiency for the NRR.

6.3 PHOTOCATALYTIC EFFICIENCY MEASUREMENTS

The photocatalytic NRR measurements are conducted in a single-compartment cell using various photoelectrodes comprised of indium tin oxide (ITO) supported nanoparticles immersed in N_2-saturated DI water with magnetic stirring under constant N_2 bubbling at the specific flow rate (e.g., 20 mL min^{-1}). Nitrogen (Ar gas for control experiments) gas was bubbled through the cell for 1 h before starting the photochemical measurements to remove dissolved oxygen gas. The air mass (AM) 1.5-irradiance of 100 mW cm^{-2} was provided by a 300 W Xe light source.

The illumination intensity near the sample surface is calibrated using a standard Si-solar cell, and the distance was adjusted to achieve 1 sun illumination. The cell was placed in the water bath to maintain the reaction temperature constant during the experiment to 20 °C and reduce thermal effects on the photocatalytic rate. The STA efficiency (η_{STA}) is calculated according to Equation 6.4:

$$\eta_{STA}(\%) = \frac{\Delta G \text{ for ammonia generation}\left(J\,mol^{-1}\right) \times \text{ammonia generated}\left(mol\right)}{\text{input light energy}\left(J\,s^{-1}\right) \times \text{reaction time}\left(s\right)} \times 100 \quad (6.4)$$

The free energy for NH_3 generation is 339 kJ mol^{-1} and the input light energy is 1000 J s^{-1}. For example, using Ag_2O–Au-685 photoelectrode for 2 h illumination under 1 sun, the amount of ammonia generated is 61.28 mg m^{-2}. The energy of ammonia produced is calculated by:

$$E_{NH_3} = \frac{61.28\,mg\,m^{-2} \times 0.0001 m^2}{17{,}000\,mg\,mol^{-1}} \times 339{,}000\,J\,mol^{-1} = 0.1222\,J \quad (6.5)$$

where the 339,000 J mol^{-1} is the Gibbs free energy of ammonia formation.

The total light energy input (one sun) is given by:

$$E_{light} = 1000\,W\,m^{-2} \times 0.0001\,m^2 \times 2 \times 3600 = 720\,J \quad (6.6)$$

Therefore, the η_{STA} is determined by:

$$\eta_{STA}(\%) = \frac{0.1222}{720} \times 100 = 0.017\% \quad (6.7)$$

By using a cutoff filter to provide visible light wavelength (400 nm < λ < 800 nm), the ammonia production rate of 47.8 mg m^{-2} is obtained after 2 h visible light illumination. In addition, the visible light intensity is measured to be 455.2 W m^{-2}. By following the above calculation, the η_{STA} is calculated to be 0.029%.

$$\eta_{STA-Vis} = \frac{\dfrac{47.8\,mg\,m^{-2} \times 0.0001\,m^2}{17{,}000\ mg\ mol^{-1}} \times 339{,}000\,J\,mol^{-1}}{455.2\ W\ m^{-2} \times 0.0001\ m^{-2} \times 2 \times 3600} \times 100 = 0.029\% \tag{6.8}$$

The apparent quantum efficiency (QE) of the NH_3 production using 685 nm band-pass filter with 15 nm full width half maximum (FWHM) is calculated as follows:

$$QE(\%) = \frac{NH_3\ generated(mol) \times 3}{the\ number\ of\ incident\ photons(mol)} \times 100 \tag{6.9}$$

The light intensity is measured to be 14 W m^{-2}, and NH_3 yield is measured after 2 h monochromatic light irradiation at 685 nm. Equation 6.10 determines the energy of each photon:

$$E = \frac{hC}{\lambda} = \frac{6.626 \times 10^{-34} J.s \times 3 \times 10^{8} m.s^{-1}}{685 \times 10^{-9} m} = 2.902 \times 10^{-19}\ J\ /\ photon \tag{6.10}$$

The local power densities at the distance of 6 cm from the light source under one sun illumination and with the monochromatic light irradiation at 685 nm with 15 nm bandwidth at FWHM are 1000 and 14 W m^{-2}, which are determined by the standard Si-solar cell.

The active surface area of the photoelectrode is 1 cm^2. Then, the energy is determined by:

$$E = 14\ Wm^{-2} \times 0.0001\ m^2 = 0.0014\ J/s \tag{6.11}$$

The number of photons per second is determined by dividing (5) by (4):

The number of photons per second:

$$n = \frac{0.0014\,J/s}{2.902 \times 10^{-19}\ J/photon} = 4.8243 \times 10^{15}\ photon/s \tag{6.12}$$

The total number of incident photons under 2 h illumination is given by:

$$N = 4.8243 \times \frac{10^{15}\ photons}{s} \times 2 \times 3600\,s = 3.4735 \times 10^{19}\ photons \tag{6.13}$$

The number of moles of photons is then calculated by:

$$n = \frac{3.4735 \times 10^{19}}{6.022 \times 10^{23}} = 5.768 \times 10^{-5}\ moles\ of\ photons \tag{6.14}$$

The ammonia yield after 2 h illumination is obtained to be 39.11 mg m^{-2}. The number of generated

$$NH_3(mol) = \frac{39.11\,mg\,m^{-2} \times 10^{-4}\,m^2}{17{,}000\,mg\,mol^{-1}} = 2.3 \times 10^{-7}\,moles\ of\ NH_3 \tag{6.15}$$

and finally, the QE is given by:

$$QE = \frac{2.3 \times 10^{-7}\,mol \times 3}{5.768 \times 10^{-5}} \times 100 \cong 1.2\% \tag{6.16}$$

The bandgap energies (E_g) of Ag_2O-Au nanocages and Ag_2O nanocubes are determined based on the Kubelka–Munk theory. The absorption spectra collected by the UV-vis spectrophotometer is multiplied by the energy and raised to the power of the inverse bandgap transition exponent ($(\alpha h\nu)^{1/p}$). α is the absorption coefficient, p is the bandgap transition dependent exponent ($p = 0.5$ for Ag_2O), and $h\nu$ is the energy per photon, calculated according to Equation 6.17:

$$h\nu = \frac{hC}{\lambda} \tag{6.17}$$

where h is Plank's constant (~4.135 × 10^{-15} eV.s), C is the speed of light in a vacuum (~3×10^{17} nm s^{-1}), and λ is the wavelength (nm). To calculate the bandgap energy, $(\alpha h\nu)^{1/p}$ is plotted against $h\nu$, known as a Tauc plot. The bandgap energy is obtained by drawing an inflection tangent in the linear region in the Tauc plot and its intersection on the energy axis where the linear portion crosses the x-axis. The position of the valence band maximum (E_{VBM}) is determined by the summation of the valence band offset at low binding energy (E_F–E_{VBM}) and secondary electron onset, which is referenced to the 21.21 eV helium source energy (work function, E_{VAC}–E_F); these are obtained by performing ultraviolet photoelectron spectroscopy (UPS). For example, for the case of Ag_2O in

$$(E_{VAC} - E_F) + (E_F - E_{VBM}) = (21.21 - 17.04) + (1.9 - 0) = 6.07\ eV \tag{6.18}$$

This energy level is converted to potential vs. RHE, assuming 4.5 eV at 0 V vs. RHE

$$E_{VBM} = 6.07 - 4.5 = 1.57\ V\ vs.\ RHE \tag{6.19}$$

6.4 PHOTOCATALYTIC NRR ACTIVITIES OF HYBRID PHOTOCATALYSTS

Ag_2O-Au nanocages with various LSPR peak positions (i.e., 655, 685, and 715 nm) are prepared by increasing the amount of Au^{3+} ions added to the AgNCs template (Figure 6.3a). As Ag atoms are etched and replaced by Au atoms, the LSPR peak position redshifts and the pore size at the walls and corners of the nanocages increases (Figure 6.3b–d). High-resolution transmission electron microscopy (HRTEM) and the Fast Fourier Transform (FFT) of 100 oriented Ag_2O-Au nanocages reveals that Ag_2O has the same orientation as Au (Figure 6.3e). Furthermore, the measurement of lattice spacings of Ag_2O (Ag–O atomic spacing, 2.23 Å) and Au (Au–Au atomic spacing, 2.68 Å) in the Ag_2O–Au-685 nanoparticles obtained from HRTEM confirms the formation of Ag_2O after oxygen treatment; these measurements are also comparable with previously reported data obtained by extended x-ray absorption fine structure (EXAFS) (Figure 6.3f).[239,240] X-ray photoelectron spectroscopy (XPS) measurements investigate the surface elemental composition and the chemical states of the as-prepared Ag_2O, Ag–Au, and Ag_2O–Au nanoparticles. The XPS survey spectra reveal that Ag_2O-Au and Ag–Au nanocages contain Ag 3d, Au 4f, and O 1 s, while Ag 3d and O 1 s are observed for Ag_2O (Figure 6.4a). A pair of spin-orbit doublets with energy peaks at 84.0 eV and 87.7 eV correspond to Au $4f_{7/2}$ and Au $4f_{5/2}$, indicating the existence of Au^o (Au-Au bonding) in Ag_2O-Au and Ag–Au nanocages, while as expected, no Au bonding energy peaks are observed for Ag_2O from the high-resolution XPS spectra of Au 4f (Figure 6.4b).[112] The doublet peaks at 368.0 eV and 374.0 eV are assigned to the Ag $3d_{5/2}$ and Ag $3d_{3/2}$, for Ag_2O-Au and Ag_2O, while a slight shift to high binding energy (0.1 eV) is observed for Ag–Au, which distinguishes between Ag^o (Ag-Ag bonding) and Ag^+ (Ag–O bonding) (Figure 6.4c).[129] The O 1 s profiles for Ag_2O and Ag_2O–Au are deconvoluted into two peaks centered at 531.5 eV and 532.6 eV (with 0.1 eV negative shift to lower binding energy for Ag_2O–Au), which are attributed to the lattice oxygen atoms of Ag_2O and the chemisorbed oxygen caused by the external –OH group or the water molecule adsorbed on the surface (Figure 6.4d).[129,130] The latter peak is also observed in the O 1 s profile for Ag–Au nanocages.

The electronic band structures of Ag_2O and Ag_2O–Au-685 nanocages are determined by optical band gaps and UPS results (Figure 6.5a). The summation of the valence band offset at low binding energy (E_F–E_{VBM}) and the secondary electron onset, referenced to the 21.21 eV helium source energy (work function (∅), E_{VAC}–E_F), provides the position of the valence band maximum (E_{VBM}) (Figure 6.6a and 6.6b). The Fermi level of Au has been reported to be 5.1 eV.[233] The optical band gap is calculated by measuring UV-vis spectra of the photocatalysts. The UV-vis spectra are analyzed by the Kubelka–Munk (KM) theory, where an inflection tangent is drawn in the linear portion of the KM function versus the energy of light absorbed (Figure 6.6c and 6.6d).[241–243] The band gap energies (E_g) of Ag_2O and Ag_2O-Au are calculated to be 2.34 and 1.6 eV, suggesting that both photocatalysts are visible-light responsive. It is important to note that the modified band gap of 1.6 eV corresponds to the hybrid Ag_2O–Au after structural and chemical modifications of AgNCs and Ag–Au nanocages. The E_{VBM} levels of Ag_2O (1.57 V) and Ag_2O–Au (1.47 V) are more positive than 1.23 V vs. RHE, indicating that both photocatalysts can drive water oxidation (Equation 6.1) through photo-generated holes, a necessary reaction to provide protons for NRR (Equation 6.2). In addition, the energy levels of the conduction band minimum (E_{CBM}) for Ag_2O (−0.77 V) and Ag_2O-Au (−0.13 V) are more negative than the NRR potential (0.05 V vs. RHE), suggesting that direct photocatalytic

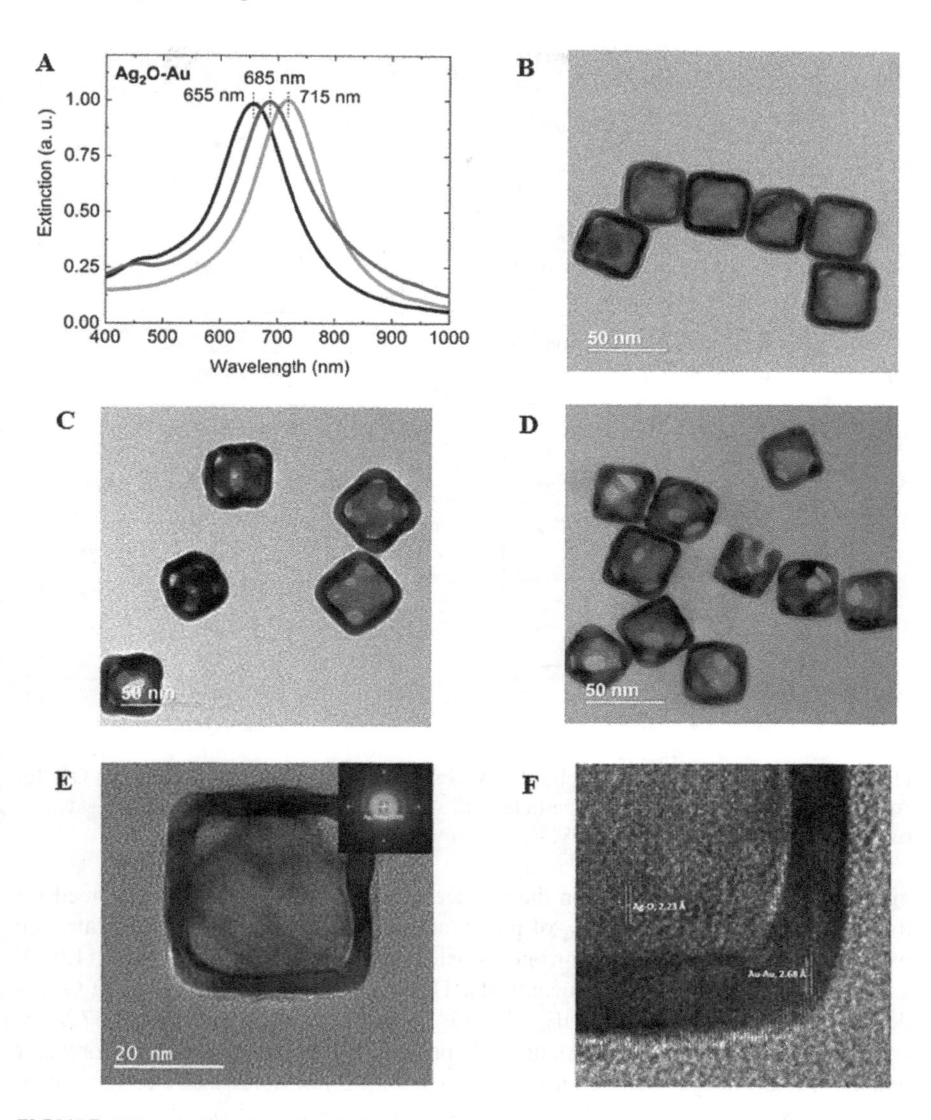

FIGURE 6.3 (a) UV-vis extinction spectra of Ag_2O-Au nanocages with various LSPR peak values. (b–d) TEM images of Ag_2O-Au nanocages with LSPR peak values at 655 nm, 685 nm, and 715 nm, respectively. As the LSPR peak redshifts from 655 nm to 715 nm, the pore size at the wall and corner of the nanocages increases. (e) HR-TEM image of Ag_2O–Au with the LSPR peak value at 685 nm, the inset shows the FFT of the nanoparticle, confirming the formation of Ag_2O after oxygen treatment. (f) Lattice spacing of Ag_2O (2.23 Å) and Au (2.68 Å) in the Ag_2O–Au-685 nanoparticle obtained from HRTEM. Reprinted with permission from Ref.[184]

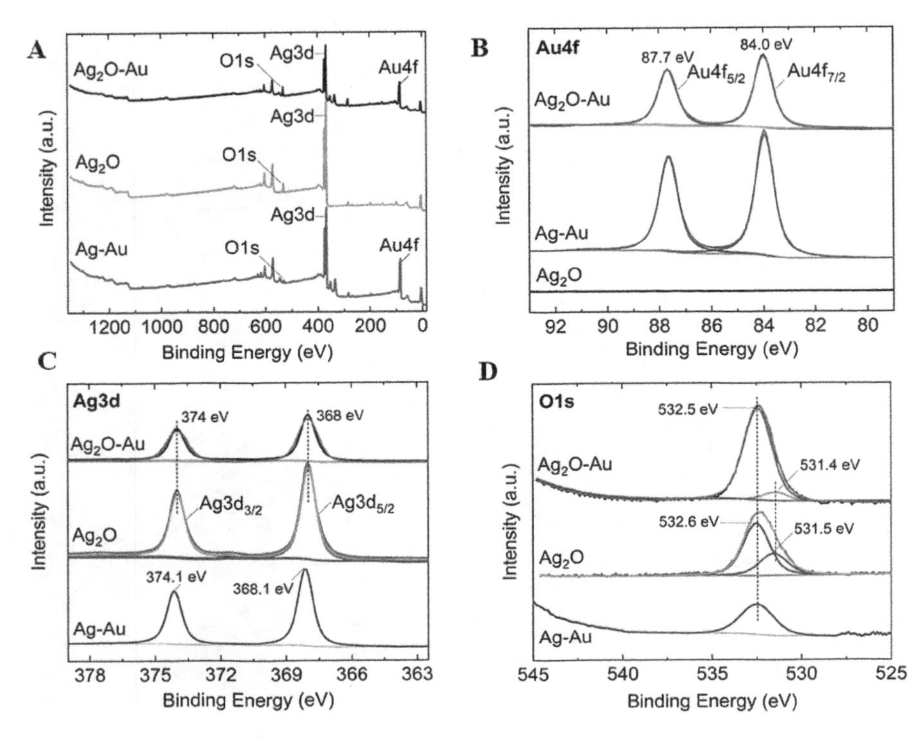

FIGURE 6.4 (a) XPS survey spectra, (b) XPS spectra of Au 4f, (c) Ag 3d, and (d) O1s for Ag_2O, Ag_2O–Au, and Ag–Au nanoparticles. All spectra were shift corrected using a standard reference C1s, C–C peak at 284.8 eV. Reprinted with permission from Ref.[184]

nitrogen reduction is possible on the surface of photocatalysts without the need for the sacrificial reagent. The E_{VBM} of photocatalysts is not favorable for the nitrogen oxidation reaction (N_2/NO), as nitrogen's oxidation potential is more positive (1.68 V vs. RHE) than the E_{VBM} of photocatalysts. The flat band potential (E_{fb}) of Ag_2O was determined to be 1.31 V vs. RHE, which is close to the E_{VBM} of Ag_2O (1.57 V vs. RHE) as shown in Figure 6.5a. Au not only provides active catalytic centers for water oxidation and nitrogen reduction through generating hot holes and electrons upon LSPR excitation, but also collects photo-generated electrons from the Ag_2O semiconductor, retards electron-hole recombination and enhances charge transfer efficiency for NRR. The photochemical nitrogen fixation tests are carried out in a single-compartment cell using photoelectrodes comprised of ITO supported nanoparticles immersed in N_2-saturated DI water with magnetic stirring under constant N_2 bubbling. The external light source shines through a one-inch quartz window placed in the front side of the cell (Figure 6.5b). A 300 W Xe light source provides the air mass (AM) 1.5 irradiance of 100 mW cm^{-2}. The illumination intensity near the sample surface is calibrated using a standard Si-solar cell, and the distance is adjusted to 6 cm to achieve one sun illumination. The cell is placed in the water bath to maintain the reaction temperature constant during the experiment to 20 °C and reduce thermal effects on the photocatalytic rate.

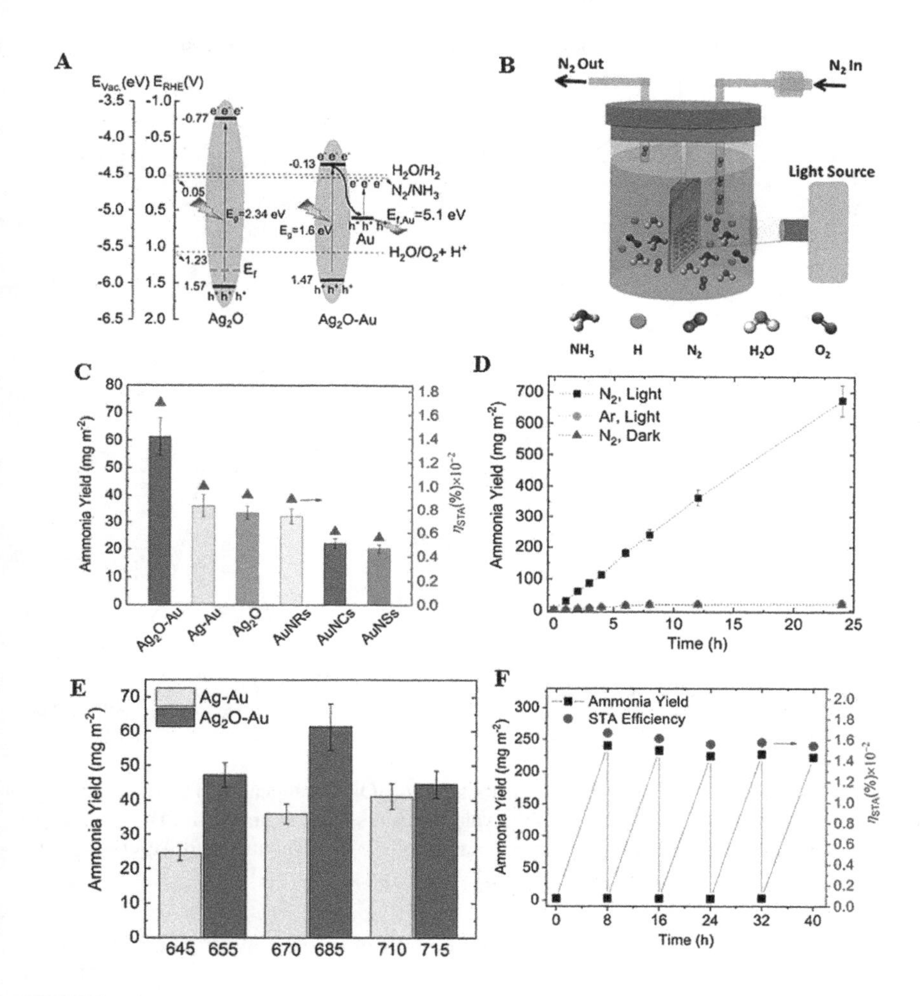

FIGURE 6.5 (a) Electronic band structure of Ag_2O and Ag_2O-Au photocatalysts for NRR. (b) Schematic of photochemical cell for NRR. (c) Ammonia yield and STA efficiency of various photocatalysts under one sun illumination for 2 h in a N_2 saturated pure water. (d) Ammonia yield of Ag_2O–Au-685 under various operating conditions over 24 h. (e) Ammonia yield for photocatalysts with various LSPR peak values before (Ag–Au) and after (Ag_2O-Au) oxygen treatment. (f) Stability test of Ag_2O–Au-685 photoelectrode for ammonia production. Five consecutive tests are carried out for the period of 8 h (total 40 h) using Ag_2O–Au-685 photoelectrode. The electrolyte solution was replaced before starting each cycle. Reprinted with permission from Ref.[184]

Among all photoelectrodes tested (i.e., Ag_2O-Au, Ag–Au, Ag_2O, Au), Ag_2O–Au-685 nanocages showed the highest activity, with the ammonia yield of 61.3 mg m^{-2} and STA of 0.017% after 2 h illumination (Figure 6.5c). This high activity is attributed to the combined effects of generated hot electrons by Au nanoparticles and photo-generated electrons by Ag_2O semiconductor that enhanced the photocatalytic

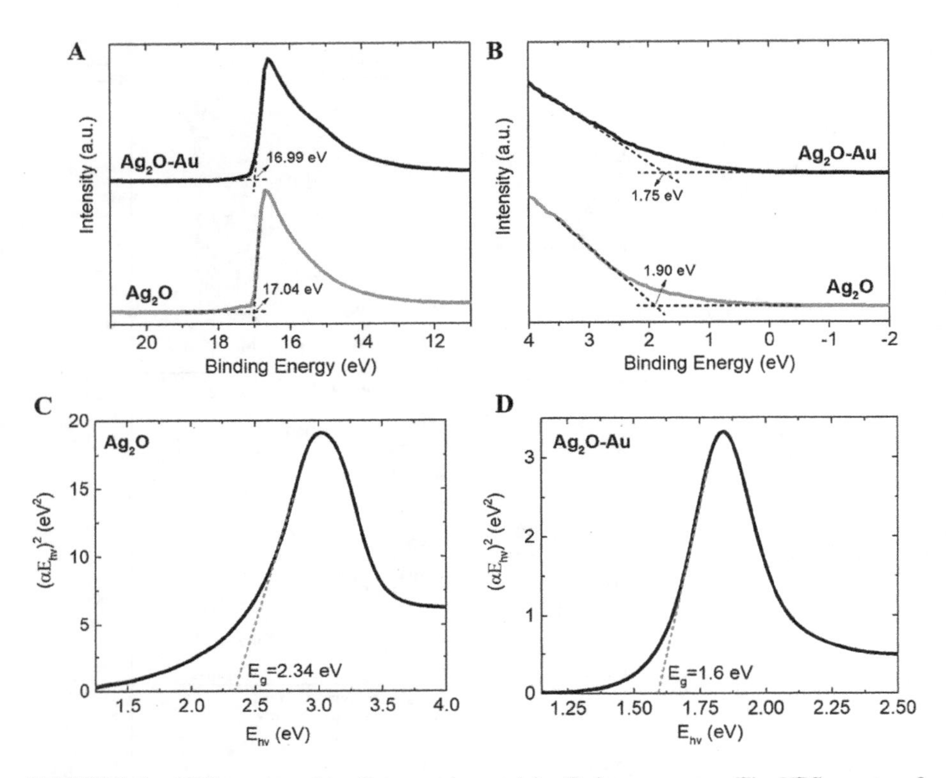

FIGURE 6.6 UPS spectra of Ag_2O nanocubes and Ag_2O-Au nanocages. The UPS spectra of (a) the secondary electron edge and (b) the valance bands are measured with He_I (21.22 eV) source radiation. (c and d) Bandgap determination of Ag_2O nanocubes and Ag_2O–Au nanocages using Kubelka–Munk theory. Reprinted with permission from Ref.[184]

NRR activity. The photocatalytic NH_3 production is further investigated by using a cutoff filter to provide visible light irradiation (400 nm $< \lambda <$ 800 nm). The NH_3 production rate of 47.8 mg m^{-2} is obtained after 2 h visible light illumination with the visible light intensity of 455.2 W m^{-2}, resulting to a $\eta_{STA\text{-}Vis}$ of 0.029%. The Ag–Au nanocages (ammonia yield: 36 mg m^{-2}, η_{STA}: 0.01%) exhibit a higher NRR activity than Ag_2O (ammonia yield: 33.5 mg m^{-2}, η_{STA}: 0.0093%). Although it was shown that Ag is not an active site for NRR, Ag–Au nanocages benefit from the "cage effect" and higher active surface area compared with solid Ag_2O nanocubes, which have higher light absorption capability.[184] It should be noted that among all solid Au plasmonic nanoparticles (i.e., nanorods (AuNRs), nanocubes (AuNCs), and nanospheres (AuNSs)), AuNRs reveal the highest NRR activity (ammonia yield: 33.5 mg m^{-2}, η_{STA}: 0.0089%). This is attributed to the two plasmon peaks (transverse (510 nm) and longitudinal (746 nm) modes) observed for AuNRs which brings advantage in light absorption compared with one plasmon peak of AuNCs or AuNSs when all other operating conditions are the same (Figure 6.5c). In addition, through the oxygen treatment of AgNCs, solid Ag_2O nanocubes are obtained. Remarkably higher NRR activity is achieved through hybrid nanocages oppose to pure Au nanoparticles.

Ag_2O–Au-685 nanocages are further tested throughout 24 h illumination. The ammonia yield rate of 28.2 mg m^{-2} h^{-1} is obtained with N_2 under illumination (N_2, light) (Figure 6.5d). Significantly lower ammonia yield rates are achieved in Ar-saturated water with illumination (0.98 mg m^{-2} h^{-1}) and N_2-saturated water without illumination (0.83 mg m^{-2} h^{-1}), indicating that N_2 is the only source of NH_3 obtained by photocatalytic NRR (Figure 6.5d). The small amounts of ammonia measured in control experiments could be attributed to the interaction of the adsorbed N_2 at the catalyst surface with water and the leakage of N_2 from the atmosphere into the cell.[158,219] Ag–Au nanocages with various LSPR peak values (i.e., 645, 670, and 710 nm) and pore sizes are tested for photocatalytic NRR for 2 h illumination before (Ag–Au) and after O_2 treatment (Ag_2O–Au) (Figure 6.5e). For all nanocages, the ammonia yield increases after O_2 treatment of Ag–Au nanocages, suggesting that the formation of Ag_2O enhances the photocatalytic NRR activity for NH_3 production. In addition, among Ag–Au nanocages with various LSPR peak values before O_2 treatment, the ammonia yield increases as LSPR redshifts from 645 nm to 710 nm. This is attributed to the fact that as LSPR redshifts, Ag is replaced with Au through the galvanic replacement, and the Ag content of the nanoparticles decreases. This is favorable for increasing photocatalytic activity, as Au is an active nanocatalyst for NRR. Ag_2O–Au-685 nanocages showed the highest photocatalytic NRR activity. This is attributed to the compromise between the pore size, the active surface area of the nanoparticle, and the Ag content (transformed to Ag_2O after O_2 treatment) of Ag_2O–Au nanocages. Higher Ag content (Ag_2O–Au-655 has the highest Ag content, see Table 6.1) is beneficial, as it is transformed to Ag_2O after O_2 treatment, which provides photo-excited electrons for NRR. In addition, in order to enhance photocatalytic NRR, it is necessary to engineer the optimum pore size; this can be done by tuning the LSPR peak position to obtain the highest active surface area while reactants and products can diffuse in and out of the cavity without restrictions. An optimum Ag content and a pore size that maximizes photocatalytic NRR for NH_3 production are achieved by using Ag_2O–Au-685 nanocages (Figure 6.5e). The apparent QE (η_{AQE}) is calculated to be 1.2% for Ag_2O–Au-685 nanocages using 685 nm bandpass filter with 15 nm bandwidth at FWHM. The performance of various hybrid photocatalysts in the literature is provided in Table 6.2. The stability of Ag_2O–Au–685 for photocatalytic NH_3 production is evaluated for 40 h by conducting five consecutive cycles, each for 8 h (assuming 8 h of daylight per day) (Figure 6.5f). The photocatalyst could maintain continuous NH_3 formation with a stable NH_3 yield and η_{STA} (92.5% performance retention). In addition, the TEM images and XPS spectra of the photocatalyst before and after the durability test show that the morphology and chemical states of nanoparticles are reasonably maintained after a 40 h photochemical test (Figure 6.7). The transient photocurrent responses are measured for Ag_2O–Au nanocages with various LSPR peak values supported on the ITO substrate, in N_2-saturated 0.5 M $LiClO_4$ (aq.) solution to evaluate the interfacial charge kinetics (Figure 6.8a). The schematic of the photoelectrochemical setup is shown in Figure 6.9. Ag_2O–Au–685 exhibits the highest photocurrent response, suggesting the efficient charge separation and transfer process using this photocatalyst. This result is in line with the greater photocatalytic activity of NRR for NH_3 production using Ag_2O–Au-685 nanocages. Furthermore, the photocurrent responses are compared in Ar- and N_2-saturated electrolyte using an

Table 6.1
Au, Ag Concentrations, Au Content (Mass % and at. %) of Various Type of Nanoparticles (Solid vs. Hollow) and their Corresponding LSPR and Transient Photocurrent Response. Au and Ag Concentrations are Determined using Inductively Coupled Plasma Emission Spectroscopy (ICPES).[b] Atomic Content (at. %) is Calculated using Au and Ag Concentrations Divided by the Molar Mass of Au (196.97 g mol^{-1}) and Ag (107.87 g mol^{-1})

Photocatalyst	Type of Nanoparticle	LSPR (nm)	Au Conc. (μg mL^{-1})	Ag Conc. (μg mL^{-1})	Au Content (mass %)	Au Content (atom %)	Photo Current (nA cm^{-2})
Ag–Au	Hollow	645	1.4	3.1	31.2	19.9	33.7
Ag_2O–Au	Hollow	655					40.2
Ag–Au	Hollow	670	2.0	3.43	37.0	24.2	36.3
Ag_2O–Au	Hollow	685					53.8
Ag–Au	Hollow	710	4.45	3.97	47.1	32.8	24.0
Ag_2O–Au	Hollow	715					33.6
Ag_2O	Solid	430	0	0.7	0	0	36.1
Au	Solid NC	535	1.2	0	100	100	11.2
Au	Solid NS	530	0.95	0	100	100	10.1
Au	Solid NR	510, 746	0.5	0	100	100	19.7

Reprinted with permission from Ref.[184]

Ag_2O–Au-685 photoelectrode (Figure 6.8b). The higher photocurrent response in Ar-saturated electrolyte (64.3 nA cm^{-2}), compared with N_2-saturated electrolyte (54.1 nA cm^{-2}), is attributed to the two possible pathways for photo-induced electrons in an N_2 atmosphere, where they can either be transferred to the ITO surface (photocurrent response) or to the adsorbed N_2 molecules at the electrode-electrolyte interface. This decreases the number of electrons transferred to the ITO and therefore lowers the photocurrent response in an N_2 atmosphere compared with an Ar atmosphere, where only one electron pathway from the photocatalyst to the ITO surface is feasible (Figure 6.8b).[242,244] The values of the transient photocurrent responses of all nanoparticles are provided in the supporting information (Table 6.1). It is important to note that ammonia yield does not change dramatically under visible light illumination (47.8 mg m^{-2}) and on-plasmon resonance excitation at 685 nm (39.11 mg m^{-2}). The difference between these two values (8.69 mg m^{-2}) is the sole contribution of the Ag_2O semiconductor, which is excited by off-plasmon resonance but has sufficient photon energy to stimulate the Ag_2O. This strongly signifies the plasmonic excitation effect, which is further promoted by charge separation at the plasmonic metal–semiconductor interface. Photoluminescence (PL) measurements are performed to examine the charge transfer for Ag_2O and Ag_2O–Au-685 photocatalysts, and emissions are measured using 375 nm excitation wavelength (Figure 6.8c). The emission intensity in the PL spectrum of Ag_2O–Au is remarkably lower than that of Ag_2O, which is attributed to the fluorescence quenching by the Au nanoparticles. Furthermore, Nyquist plots are obtained by performing electrochemical impedance spectroscopy (EIS) under illumination (Figure 6.8d). The charge transfer resistance

Table 6.2
Summary of Prior Studies on Photocatalytic Nitrogen Fixation with Various Hybrid Photocatalysts

Photocatalyst	Reaction Conditions	Light Source	Sacrificial Reagent	NH_3 Yield Rate (nmol cm^{-2} h^{-1})[a] or (nmol mg^{-1} h^{-1})[b]	η_{STA} (%)	η_{AQE} (%)	Ref.
Au/black Si/Cr	10 mL H_2O (l), ambient conditions	300 W, Xe lamp, 200 mW cm^{-2}	Sodium sulfite	78.3[a]	0.0037[d]	0.003 at 500 nm	160
TiO_2/Au/a-TiO_2	15 mL H_2O (l), ambient conditions	300 W, Xe lamp, 100 mW cm^{-2}	No	13.4[a]	0.0013[d]	0.005 at 254 nm	161
Au/Nb-$SrTiO_3$/Ru	215 μL H_2O + HCl (l), ambient conditions	Xe lamp, 550–800 nm	Ethanol	1.0[a]	NA	3.8×10^{-5} at 630 nm	163
Au/Nb-$SrTiO_3$/Zr/ZrO_x	H_2O + HCl (l), ambient conditions	Xe lamp, 550–800 nm	Ethanol	6.5[a]	NA	0.95 at 600 nm	117
TiO_2 (Rutile)	200 mg catalyst in 200 mL water	High pressure Hg lamp 300 W	No	8.8[b]	0.02	0.7 at $\lambda <$ 350 nm	219
m-PCN-V (98)	200 mg catalyst in 100 mL water	Xe lamp	No	14.58[b]	0.1	~1.0 at $\lambda <$ 420 nm	245
BOB-001-OV	50 mg catalyst in 100 mL water	300 W Xe lamp with 420 nm cutoff filter	No	104.2[b]	NA	0.23 at 420 nm	227
Au/TiO_2–OV[c]	100 mg catalyst in 72 mL water and 8 mL methanol	300 W Xe lamp with 420 nm cutoff filter	Methanol	78.6[b]	NA	0.82 at 550 nm	236
1 mol% Mo-doped $W_{18}O_{49}$	10 mg of catalyst in 10 mL water	300 Xe lamp, 200 mW cm^{-2}	Sodium sulfite	195.5[b]	0.028	0.33 at 400 nm	244
Au–$Ru_{0.31}$	0.2 mg of catalyst in 3 mL water	300 Xe lamp, 400 mW cm^{-2}	No	101.4[b]	NA	0.21 at 350 nm	234
Au–Ag_2O nanocages	40 mL H_2O (l), ambient conditions	300 W, Xe lamp, 100 mW cm^{-2}	No	165.9[a]	0.017	1.2 at 685 nm	184

Reprinted with permission from Ref.[184]

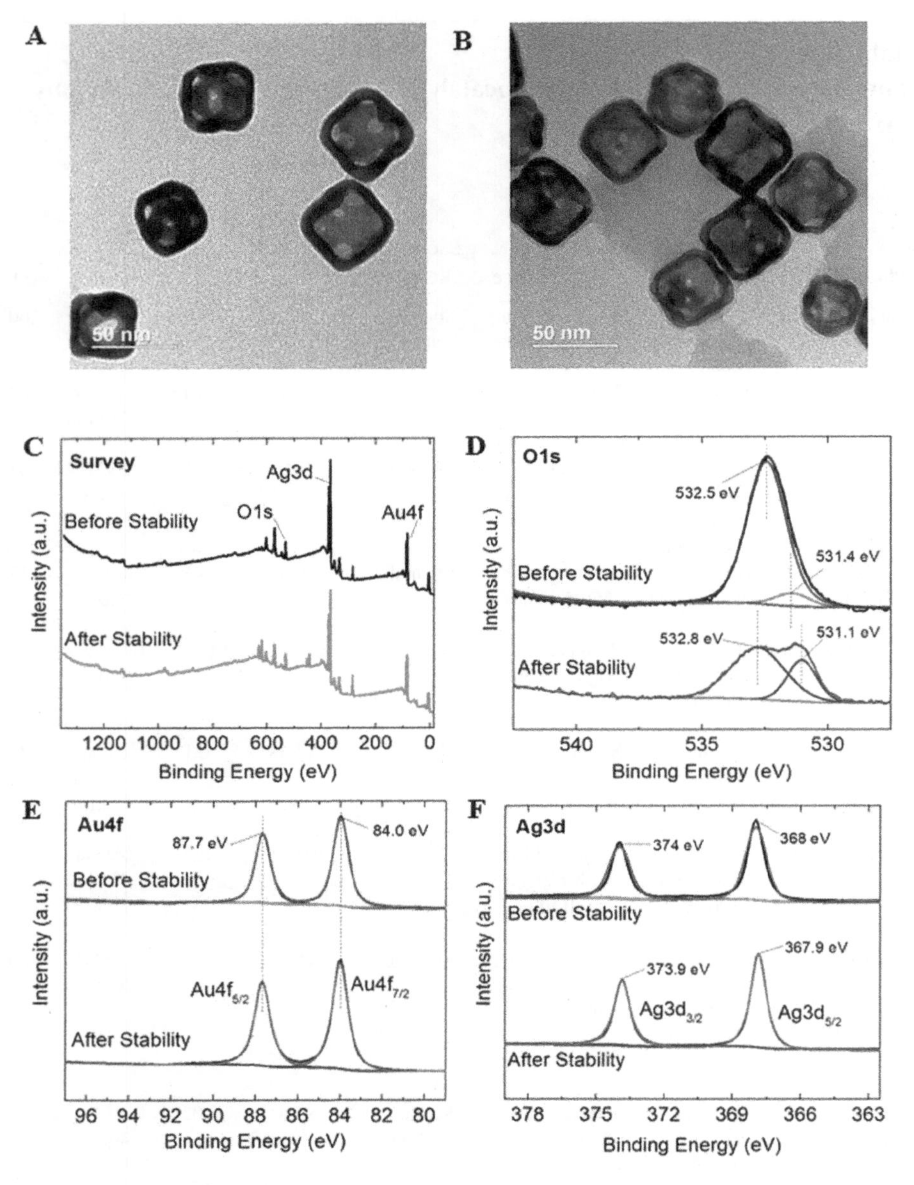

FIGURE 6.7 TEM images of Ag_2O–Au-685 nanocages (a) before and (b) after the stability test in the photocatalytic reaction. XPS spectra of Ag_2O–Au-685 photocatalyst before and after the stability test. (c) XPS survey spectra, (d) XPS spectra of O 1 s, (e) Au 4f, and (f) Ag 3d. All spectra were shift corrected using a standard reference C1s, C–C peak at 284.8 eV. Reprinted with permission from Ref.[184]

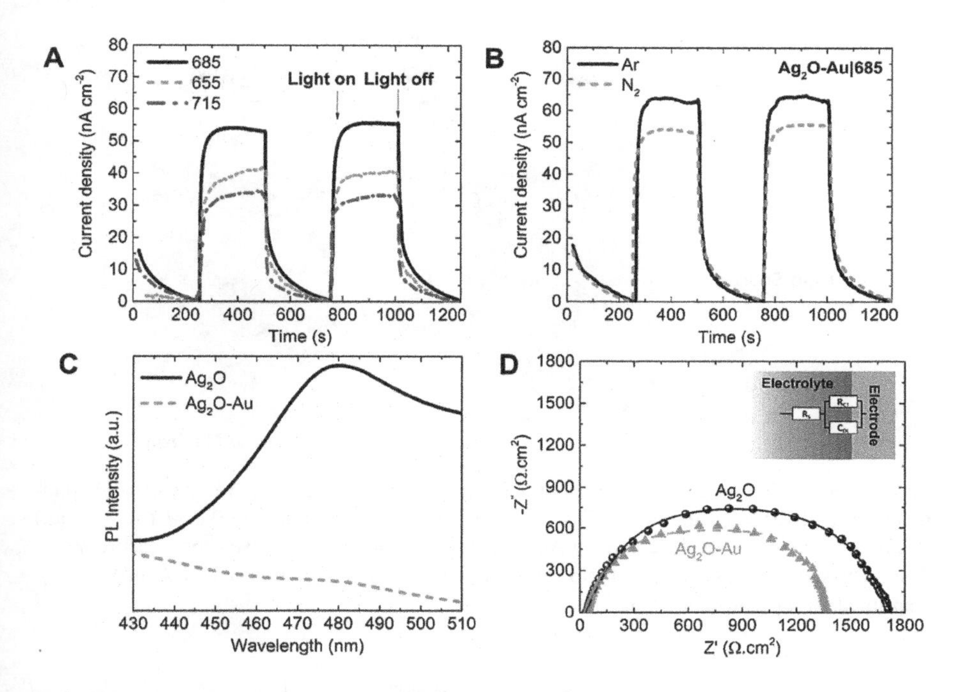

FIGURE 6.8 (a) Transient photocurrent response of Ag_2O-Au with various LSPR peak values. (b) Transient photocurrent response of Ag_2O–Au-685 in N_2 and Ar-saturated 0.5 M $LiClO_4$ (aq.) solution under 1 sun illumination. Photocurrent responses are measured at the open circuit voltage of the cell. (c) Photoluminescence spectra of Ag_2O and Ag_2O–Au-685 photocatalysts. Both samples were excited with 375 nm excitation source. (d) Nyquist plots of the Ag_2O–Au-685 under photo-irradiation at an applied potential of 0.5 V (vs Ag/AgCl) in 0.5 M $LiClO_4$ (aq.) solution. The equivalent circuit model in the inset comprises charge transfer resistance ($\boldsymbol{R}_{CT}$), double layer capacitance ($\boldsymbol{C}_{DL}$), and ohmic resistance ($\boldsymbol{R}_S$). $\boldsymbol{R}_{CT}$ was obtained from the diameter of semicircles for each photocatalyst. Reprinted with permission from Ref.[184]

(R_{CT}) of Ag_2O–Au (1.34 kΩ.cm^2) is smaller than that of Ag_2O (1.68 kΩ.cm^2), suggesting an enhanced transfer of photo-generated charge carriers, which could promote an efficient photocatalytic NRR. Isotopic labeling experiments using $^{15}N_2$ gas further confirm the source of NH_3 formation in photochemical NRR tests. The amounts of ammonia produced with $^{14}N_2$ and $^{15}N_2$ after photo-irradiation for 4 h are 16.5 μM and 17.1 μM, which were analyzed using the indophenol method. Detailed discussion on various techniques for measuring ammonia accurately and essential control experiments will be discussed in Chapter 7. The amount of ammonia is further measured by the NMR method. ^1H NMR spectrum obtained from the sample in the photochemical $^{14}N_2$ reduction experiment lie at a chemical shift of triplet coupling of $^{14}N_2$ similar to that of standard $^{14}NH_4^+$ samples (J-coupling: 52 Hz). Doublet coupling of $^{15}N_2$ (J-coupling: 73 Hz) is obtained after photochemical $^{15}N_2$ reduction reaction which agrees well with the standard $^{15}NH_4^+$ solution. A small amount of $^{14}NH_4^+$ is observed (4 μM) after $^{15}N_2$ reduction experiments, indicating the possible leakage of $^{14}N_2$ from the atmosphere into the cell (Figure 6.10). By carefully isolating the cell and

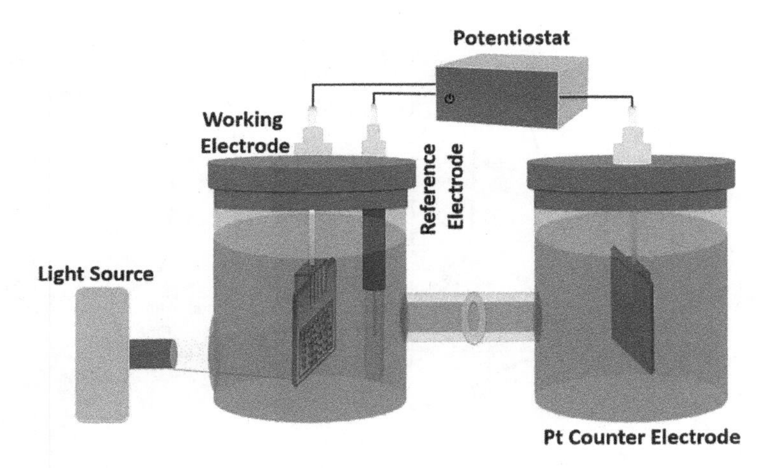

FIGURE 6.9 The schematic of the photoelectrochemical setup for performing transient photocurrent and EIS measurements. The electrolyte is N_2- or Ar-saturated 0.5 M $LiClO_4$ (aq.) solution. Pt counter electrode and Ag/AgCl reference electrode were used for measurements. The light source is 300 W Xe lamp with one sun illumination intensity (100 mW cm^{-2}). Reprinted with permission from Ref.[184]

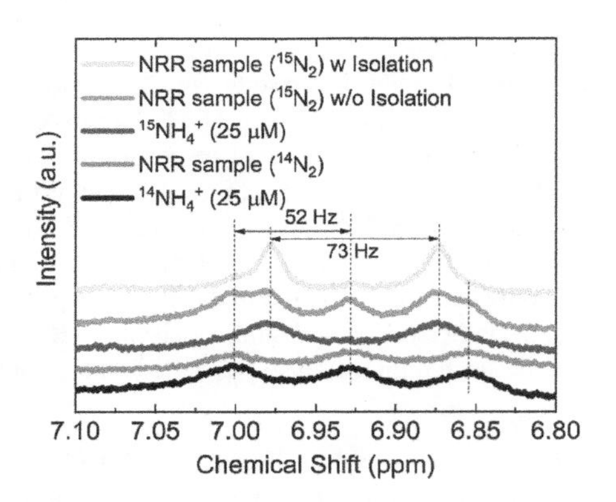

FIGURE 6.10 ^{1}H NMR spectra of samples after photochemical $^{15}N_2$ ($^{14}N_2$) reduction reaction under 1-sun illumination for 4 h using Ag_2O–Au-685 nanocages and standard $^{15}NH_4^+$ and $^{14}NH_4^+$ samples. Reprinted with permission from Ref.[184]

maintaining a slight positive pressure of $^{15}N_2$ through connecting the cell's headspace to the gas sampling bag, we successfully decreased the leakage of $^{14}N_2$ (Figure 6.10). The amount of ammonia quantified using ^{1}H NMR measurement is very close to the amount of ammonia measured using indophenol method (16.9 μM for $^{14}NH_4^+$ and 16.6 μM for $^{15}NH_4^+$ using ^{1}H NMR measurements) which further confirms that the supplied N_2 gas is the major source of NH_3 formation in the system.

The work discussed in this chapter opens up a new avenue for the design of an efficient visible-light responsive hybrid hollow plasmonic-semiconductor photoelectrode with broad applications in the photo(electro)chemical energy conversion systems.

7 Ammonia Detection

7.1 INTRODUCTION

Accurate determination of ammonia produced during electrochemical and photochemical nitrogen fixation is imperative to report ammonia formation rate, Faradaic efficiency, and energy efficiency. Ammonia is a small, polar molecule, and it is highly soluble in water (482 g L^{-1} at 24 °C).[101,146] The quantitative measurement of ammonia is typically performed in aqueous solutions, where the concentration of absorbed ammonia in the solution is determined using various techniques. However, the measurement is affected by background ammonia in the system and false-positive results for ammonia synthesis, distinguished from Faradaic reactions for ammonia synthesis at the electrode through a set of control experiments. Ammonia is an aqueous solution that exists in two forms, the ammonium ion (NH_4^+) and un-ionized ammonia (NH_3), and it follows the equilibrium relationship (Equation 7.1).[101] The relative concentrations of two forms are pH- and temperature-dependent (Equations 7.2 and 7.3).[101,246,247]

$$NH_3(aq) + H_2O(l) \leftrightarrow NH_4^+(aq) + OH^-(aq) \tag{7.1}$$

$$pK_a = 0.09018 + 2729.92(273.15 + T)^{-1} \tag{7.2}$$

Fraction of un-ionized ammonia is

$$(\%) = 100\,(1 + 10^{pKa - pH})^{-1} \tag{7.3}$$

where pKa is the dissociation constant for ammonium ion and T is the temperature (°C). Therefore, the fraction of unionized ammonia in the solution increases as the pH or the temperature of the solution increases. For the determination of total ammonia concentration, the pH of the solution is adjusted to shift the equilibrium of Equation (7.1) toward either ammonium ion or un-ionized ammonia, whose concentration is determined after that. There are various analytical methods and instruments for determining ammonia concentration, including Nessler's reagent method, indophenol blue method, salicylate method, ion chromatography, ammonia gas-sensing electrode and ammonium ion-selective electrode, fluorometric method, conductivity method, titrimetric method, and enzymatic method.[248,249] Fourier transform infrared spectroscopy (FTIR) and ^1H NMR have also been used to detect ammonia. In particular, ^1H NMR has been widely used in isotopic labeling experiments, in which ^{15}N-labeled N_2 is used as the feed gas to confirm the origin of N in the produced ammonia. Hydrazine (N_2H_4), a common byproduct of N_2 reduction reaction, is usually determined using Watt and Chrisp's method.[250] This spectrophotometric method measures the yellow color developed upon the addition of para-dimethylaminobenzaldehyde to dilute hydrochloric acid solutions containing hydrazine.

7.2 NESSLER'S REAGENT METHOD

Nessler's reagent was invented by Julius Nessler in 1856. This method consists of mercury (II) iodide and potassium iodide in an alkaline solution, either sodium hydroxide or potassium hydroxide.[251] Nessler's reagent gives a colored complex when added to solutions containing low ammonia concentrations, and a colloidal solution is formed (Equation 7.4).

$$2K_2HgI_4 + NH_3 + 3KOH \rightarrow HgO.\ Hg(NH_2)I \downarrow\ + 7KI + 2H_2O \tag{7.4}$$

As a result, the concentration of ammonia in the solution can be determined colorimetrically at wavelengths between 400 and 425 nm.[252–254] Ammonia detection using the Nessler's reagent method is interfered with by various metal cations (except sodium and potassium), hydrazine, and carbonyl compounds.[255–257] Due to the Nessler's reagent's high alkalinity, metal cations may precipitate as hydroxides, creating turbidity that interferes with colorimetric measurements. A common approach to clarify turbid samples is to add zinc sulfate and sodium hydroxide solutions followed by filtration.[258] Rochelle salt (potassium sodium tartrate tetrahydrate, $KNaC_4H_4O_6{\cdot}4H_2O$) solution is then added to eliminate residual cations' influence to prevent the formation of cloudy samples.[255] Nessler's reagent contains mercury, which is a hazardous material. Therefore, it has to be stored, handled, dispensed and disposed of properly according to laws and regulations.

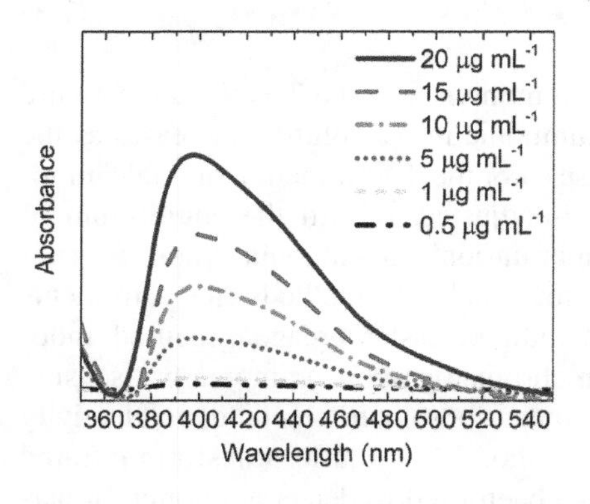

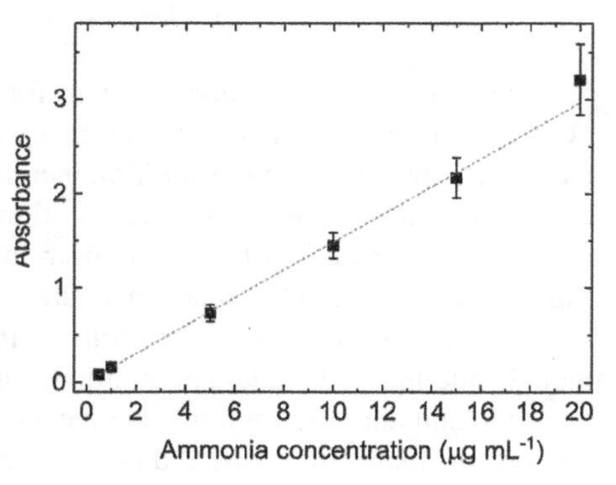

FIGURE 7.1 UV–vis absorbance spectra for standard ammonia solutions with various concentrations. UV–vis calibration curve for ammonia quantification using Nessler's method. Known concentration of ammonium ions are added to 0.5 M $LiClO_4$ (aq.) electrolyte and mixed thoroughly with 1 mL of 0.2 M $KNaC_4H_6O_6$ and 1 mL of Nessler reagent and then the absorbance at 400 nm is measured by the UV–vis spectrophotometer. The value of blank electrolyte is subtracted from all other concentrations as background. Reprinted with permission from Ref.[176] Copyright 2020, American Chemical Society.

To obtain the calibration curve in the Nessler's reagent method, a known volume of standard ammonia solution is added to the tube (Figure 7.1). Next, the tube is filled with 0.5 M of the electrolyte solution used in NRR experiments until the total volume reaches 10 mL. Then, 1 mL of 0.2 M potassium sodium tartrate in DI water is added to each of the tubes. Finally, 1 mL of Nessler reagent is added to each of the tubes and mixed thoroughly. The tubes are kept undisturbed for 20 min for color development and then the absorbance at 400 nm is measured using a spectrophotometer. The absorbance for the blank sample without adding a standard NH_3 solution is subtracted from all samples for background correction.

7.3 INDOPHENOL BLUE METHOD

The indophenol blue method is based on the Berthelot reaction. It is one of the standard methods for examining water and wastewater.[259–261] In this method, blue colored indophenol is formed from ammonia reaction with hypochlorite and phenol in an alkaline medium.[262] This reaction proceeds through several steps. The first is the reaction between ammonia and hypochlorite at pH 9.7–11.5 to give monochloramine. The monochloramine then reacts with phenol to give quinone chloramine in the presence of sodium nitroprusside, which acts as a catalyst. The quinone chloramine further reacts with phenol to form yellow associated indophenol. The indophenol dissociates in an alkaline medium to give a blue color, which can be quantitatively determined calorimetrically at wavelengths between 630 and 650 nm.[263,264]

The indophenol blue method is specific for ammonia species since organic nitrogen compounds, nitrite, nitrate, and various electrolytes only slightly interfere.[265–267] However, when their concentrations in the solution are more than ammonia, they may suppress the indophenol blue reaction.[264] In addition, interference produced by the precipitation of magnesium and calcium ions at high pH can be eliminated by complexing these ions with citrate.[268]

To obtain the calibration curve in this method, 2 mL of sample (standard ammonia solution or sample from the experiment) is added to a tube vessel. Then, 2 mL of a 1 M sodium hydroxide aqueous solution containing 5 wt% salicylic acid and 5 wt% sodium citrate is added to the tube vessel, followed by the addition of 1 mL of 0.05 M sodium hypochlorite solution and 0.2 mL of 1 wt% sodium nitroferricyanide aqueous solution. After 2 h at room temperature, the absorption spectrum was measured using a UV–vis spectrophotometer. The formation of indophenol blue was determined using the absorbance at a wavelength of 655 nm. The concentration-absorbance curves were calibrated using standard ammonia solutions (e.g., 1 μg mL^{-1}) with various dilution factors (Figure 7.2). The blank sample's absorbance without adding standard NH_3 solution is subtracted from all samples for background correction.

7.4 ION CHROMATOGRAPHY

Cationic species are separated based on their interaction with a cation-exchange material (stationary phase) and an eluent (mobile phase) in an ion chromatograph. As the eluent runs through a column packed with the cation-exchange material, cationic species move through the column at different speeds and separate from each other.

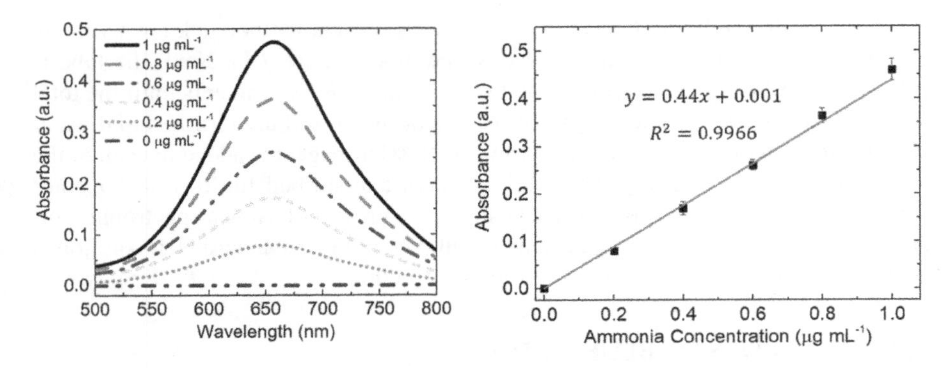

FIGURE 7.2 UV–vis absorption curves for samples with known NH_4^+ concentration using indophenol blue method. Calibration curve for NH_3 quantification based on the values of absorption peaks at 655 nm. The peak absorbance value at 655 nm is measured by UV–vis spectrophotometer. The value of blank electrolyte is subtracted from all other concentrations. Reprinted with permission from Ref.[184]

The separated cations are typically detected using a conductometric detector, and they are distinguished by comparing their retention times with those of known standard samples run individually. Quantitative measurements are conducted by measuring peak areas and comparing them to the areas obtained from standard samples.[269] Ion chromatography provides the advantage of simultaneous separation and detection of alkali and alkaline earth metal ions and ammonium ions in a single run with high sensitivity and good reproducibility.[270,271] However, ions with similar retention times may interfere with each other, especially when these ions are present in different concentrations. For example, common cation-exchange materials have similar selectivity for ammonium and sodium ions, and it is difficult to obtain a suitable resolution if the concentrations of these two cations are remarkably different.[272] To overcome this problem, it is recommended to use columns with higher cation-exchange capacities, which can provide better resolution of ammonium ions from sodium ions.[273] In addition, macrocyclic ligands can form complexes by selectively coordinating cations within their cavities and using them either in the stationary phase or in the mobile phase can significantly change the selectivity of the column toward ammonium and sodium ions.[274,275] A column-switching technique can be utilized to increase the resolution between ammonium and sodium ions.[276]

7.5 ^1H NUCLEAR MAGNETIC RESONANCE (NMR) QUANTIFICATIONS

^1H NMR has been widely used recently to measure ammonia quantitatively. ^1H NMR can distinguish the N isotopes, as the scalar interaction between ^1H and ^{15}N in $^{15}NH_4^+$ results in a splitting of the ^1H resonance into two symmetric signals with a spacing of 73 Hz, while the ^1H resonance coupled to ^{14}N in $^{14}NH_4^+$ is split into three symmetric signals with a spacing of 52 Hz.[121,145]

The $^{15}N_2$ gas was pre-purified by passing through the acid solution (1 mM H_2SO_4) followed by distilled water traps to remove any NO_x and NH_3 contamination in the $^{15}N_2$. $^{14}NH_4$ is quantified from its signal at 6.93 ppm, and a calibration curve is obtained by the integration of this signal (6.93 ppm) as a function of concentration using $^{14}NH_4^+$ standard solutions made up from $^{14}NH_4Br$. $^{15}NH_4$ is quantified from its signal at 6.97 ppm, and a calibration curve is obtained by the integration of this signal (6.97 ppm) as a function of concentration using $^{15}NH_4^+$ standard solutions made up from $^{15}NH_4Cl$ (Figure 7.3). For the aqueous electrolyte, 500 μL of the sample is mixed thoroughly with 100 μL of DMSO-d^6 as a locking solvent. For the non-aqueous electrolyte, 500 μL of a sample is mixed thoroughly with 50 μL of DMSO-d^6 and 50 μL of 0.5 M H_2SO_4 (aq.). All experiments were carried out with water suppression and 1000 scans. All peak areas were normalized to the DMSO-d^6 area peak at 2.49 ppm. This method is sensitive to detect ammonia from measurements down to 51 ppb using 800 MHz NMR instrument.

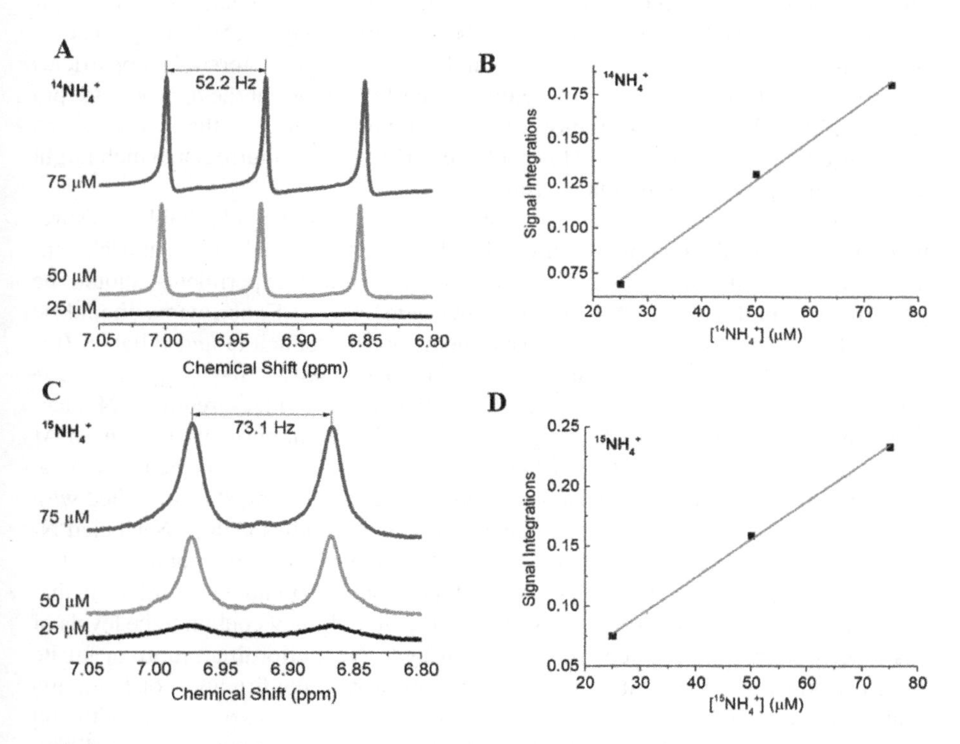

FIGURE 7.3 (a) ^1H NMR spectra for the standard $^{14}NH_4^+$ solutions with various concentrations. (b) Peak area calibration curve of the ^1H NMR signal at 6.93 ppm for standard solutions of $^{14}NH_4^+$ (25, 50, 75 μM). (c) ^1H NMR spectra for the standard $^{15}NH_4^+$ solutions with various concentrations. (d) Peak area calibration curve of the ^1H NMR signal at 6.97 ppm for standard solutions of $^{15}NH_4^+$ (25, 50, 75 μM). All peak areas were normalized to the DMSO-d^6 area peak at 2.49 ppm. Reprinted with permission from Ref.[184]

7.6 VARIOUS SOURCES OF CONTAMINATION IN NRR EXPERIMENTS

Various sources of contamination might result in false-positive ammonia measurements, which should be treated carefully. Most notably, the proton exchange membrane (e.g., Nafion) can act as the source or sink for ammonia during N_2 electrolysis. For instance, Nafion 211 and cation-exchange membrane are commonly used in the gas- and liquid-phase measurements. In the gas-phase system, the membrane is washed with dilute H_2SO_4 (aq.) and DI water before use in each measurement to extract ammonia present in the membrane. In addition, in the liquid-phase experiments, the membrane should be exchanged frequently from time to time and for each new operating condition. Recent studies suggested the use of a separator (Daramic or Celgard), which has shown a greater performance in recovering ammonia at the cathode side.[145] $^{15}N_2$ (98 atom% ^{15}N) should be purified with the acid trap to remove any NO_x or NH_x species present in the gas stream. It is noted that the reduction of NO_x to NH_3 is feasible at the catalyst surface, resulting in false-positive ammonia if the gas stream is not cleaned carefully. The use of N-containing (e.g., $AgNO_3$) reagents during the nanoparticle synthesis process should be carefully monitored. Nanoparticles must be cleaned carefully by centrifugation at least two times at the appropriate rpm (e.g., 10,000–12,000 rpm) to remove excess capping agent after the synthesis and before use in NRR experiments. This could be a new source of nitrogen, which might result in false-positive ammonia measurement.

Greenlee et al. proposed a general approach to the accurate and reliable measurement of synthesized ammonia (Figure 7.4a).[146] (i) For each catalyst and at each condition, a set of nonelectrochemical and electrochemical experiments should be performed in both Ar- and N_2-saturated environments. (ii) Experiments should be repeated to verify that ammonia measurements are reproducible and reliable. (iii) The amount of ammonia produced should be calculated as $\Delta NH_{3,N2,echem\text{-}ambient} - \Delta NH_{3,Ar,echem\text{-}ambient}$, where $\Delta NH_{3,N2,echem\text{-}ambient}$ is the ammonia measured in an N_2 electrochemical experiment and $\Delta NH_{3,Ar,echem\text{-}ambient}$ is the ammonia measured in an Ar electrochemical experiment. Both of them include background and nonelectrochemical ammonia subtractions. (iv) Once a promising N_2 reduction catalyst is identified with the general approach presented in Figure 7.4a, isotope studies using ^{15}N-labeled N_2 should be conducted to verify that N_2 is, in fact, being reduced to ammonia. The amount of ammonia measured in the isotope studies should be similar to that measured in the experiment using $^{14}N_2$. Commercial $^{15}N_2$ may contain trace levels of ^{15}N-labeled ammonia, NO_x contaminants that hydrolyze, affording nitrate or nitrite ions, and nitrous oxide[277]. It is recommended even with confirmation of ammonia synthesis with isotope studies, a full set of Ar and N_2 control experiments with and without applied potential should be performed. In addition, a combination of different ammonia detection methods can provide more reliable and convincing results compared to a single approach because the latter may suffer from interference.[88,278] Andersen et al. suggested the protocol for benchmarking electrochemical NRR (Figure 7.4b).[145] The authors suggest that adventitious ammonia and other nitrogen-containing compounds in the setup should be determined first (part 1 in Figure 7.4b).

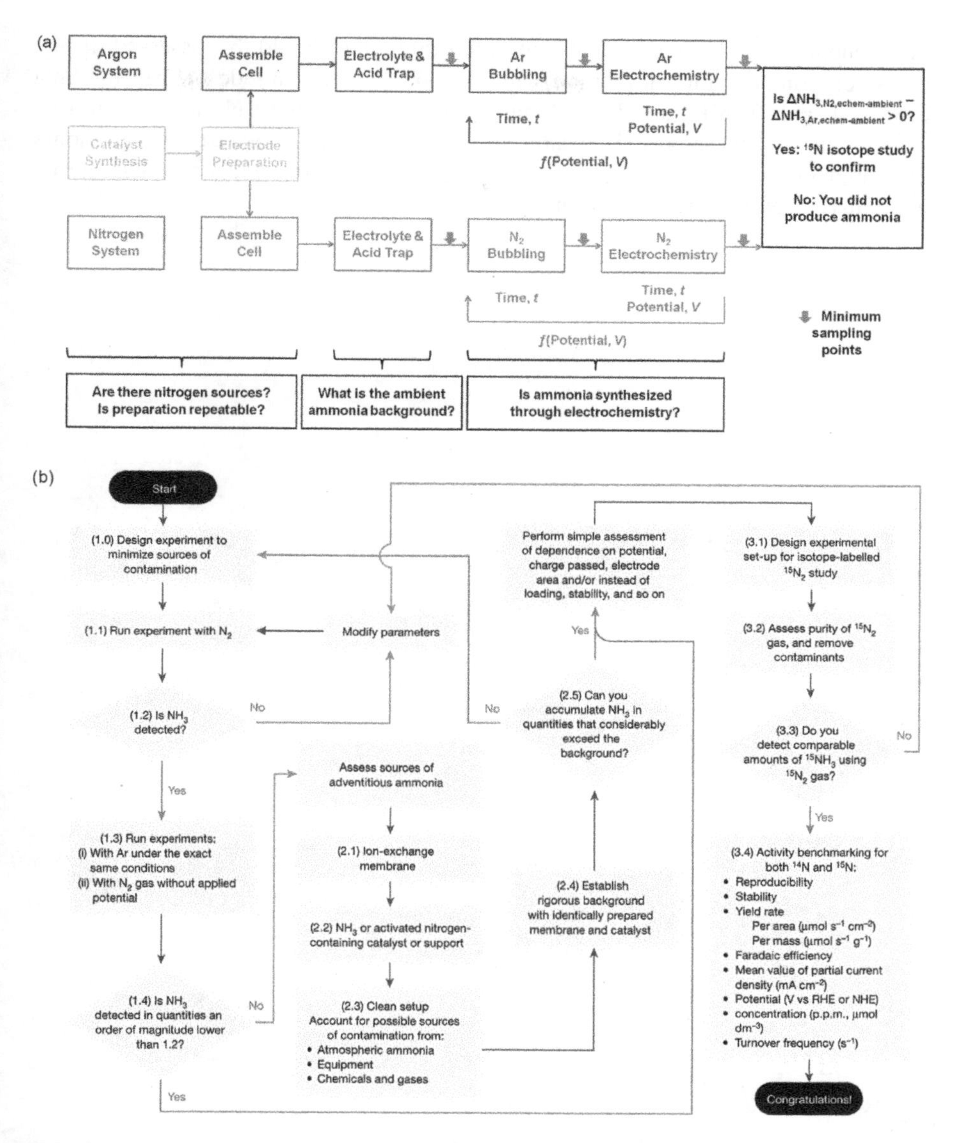

FIGURE 7.4 (a) Flow diagram of a general approach to the accurate and reliable measurement of electrochemical ammonia synthesis. Reproduced with permission from Ref.[146] Copyright 2018 American Chemical Society. (b) Suggested protocol for the benchmarking of electrochemical nitrogen reduction. The three parts in the diagram include, (i) identifying the sources of contamination in NRR experiments, (ii) membrane and catalyst contamination, and (iii) isotopic labeling experiments. Reproduced with permission from Ref.[145]

Subsequently, the sources of adventitious ammonia should be assessed if the measured contamination levels are within an order of magnitude of the ammonia produced (part 2 in Figure 7.4b). Finally, an isotope-labeled $^{15}N_2$ study should be carried out, and the synthesized ammonia should be comparable to the ammonia measured in part 1. A recent study also pinpointed the importance of gas purity, particularly $^{15}N_2$ in NRR experiments.[279]

8 Reaction Mechanisms for Nitrogen Fixation

8.1 INTRODUCTION

The N_2 reduction reaction proceeds with breaking the N≡N triple bond and forms three N−H bonds for ammonia formation. On a heterogeneous catalyst, N_2 will bond to the catalyst surface and then dissociate to two adsorbed N atoms (a dissociative mechanism). It may be possible that N_2 becomes protonated by forming N_2−H, and further hydrogenation results in creating two ammonia molecules. According to the Sabatier principle, the optimal catalyst will have the intermediate binding energy of a key reaction intermediate (e.g., NH_x^* or $N_2H_x^*$ species, where $0 \leq x \leq 2$ and * denotes a surface-bound species).

DFT calculations and microkinetic analysis revealed how this trade-off identified Fe and Ru catalysts as the optimal materials for the Haber–Bosch process. For the dissociative pathway, it was shown that the binding energies of NH_x ($0 \leq x \leq 2$) species correlate with each other on late transition metals.[280] This correlation relates the binding energy of all reaction intermediates on various catalyst surfaces using a single descriptor, which is the binding energy of N^* to the surface. Brønsted–Evans–Polanyi (BEP) relationships have been shown to hold for elementary N–N dissociation and N–H bond formation steps, linearly correlating the elementary activation barriers with the reaction energies.[281] This leaves the metal–N^* binding energy as a single descriptor that dictates all elementary reaction energies and activation barriers. Therefore, the overall rate of the ammonia synthesis reaction on late-transition-metal surfaces.

On the other hand, discovering and designing high-performance catalysts by optimizing their structure and composition requires advanced experimental techniques to evaluate, characterize, and understand photo-electrocatalysts. To this end, existing and future advances in in situ/operando characterization techniques (e.g., vibrational spectroscopy, X-ray spectroscopy) may bridge this gap in understanding and accelerating catalysts' development.

8.2 SABATIER PRINCIPLE AND SCALING RELATIONS

Sabatier principle for N_2 reduction reaction rate results in generating a "volcano" plot, with the activity is plotted against N^* binding energy (Figure 8.1a). Metal catalysts to the left (strong N^* binding) result in lower ammonia synthesis rates due to slow N–H formation (limited by product desorption), and metals to the right (weak N^* binding) are limited by N_2 activation. The BEP linear correlation between the N_2 dissociation transition state stability and the N^* binding energy is shown in Figure 8.1b. It reveals that no metal exists that combines intermediate N^* binding and a low energy N_2 dissociation transition state (bottom middle of plot). As the optimal catalyst at the top of

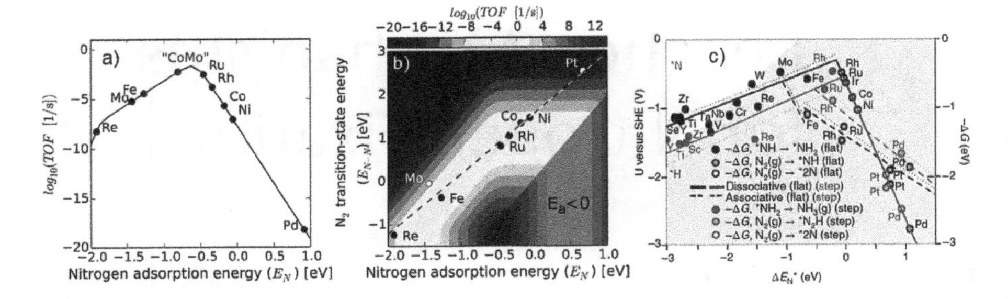

FIGURE 8.1 Computational predictions and theory-based limitations for heterogenous (electro) catalysts. (a) Volcano plot for ammonia synthesis on late transition metals. Reprinted with permission from Ref.[283] (b) Scaling relationship for N_2 dissociation transition state intermediate on late transition state metals. Reprinted with permission from Ref.[283] (c) Proposed volcano plot for electrochemical nitrogen reduction on late transition metals. Reprinted with permission from Refs.[52,280]

the 'volcano' indicates, a significant activation barrier for both N_2 dissociation and N–H bond formation exists. Therefore, the Haber–Bosch reaction must be performed at an elevated temperature (400–500 °C) to reach an acceptable rate. As higher temperature limits equilibrium conversion for the overall reaction (forward reaction $N_2 + 3H_2 \leftrightarrow 2NH_3$), high pressure must be used to reach reasonable conversions. Electrochemical NRR will be initiated by forming N_2H^* species before completely breaking the N≡N triple bond. However, the volcano relationship suggests that late transition metals would have limitations to reduce N_2 through the electrochemical path (Figure 8.1c).[280] This is due to the significant overpotential for N_2 reduction reaction regardless of the associative and dissociative pathways. Since hydrogen evolution has lower overpotentials and kinetically favorable, N_2 reduction suffers from a selectivity challenge. This is also consistent with the experimental findings, which suggest that no late-transition metal is highly active or selective for electrochemical NRR at ambient temperature. Catalytic systems/materials that either lower activation barriers relative to the late-transition metal BEP relationships or break scaling or BEP relationships altogether are needed.[52] For catalytic reactions involving multiple reaction intermediates, the scaling relations indicate that the binding energies of different intermediates are strongly correlated, limiting the possibilities of finding good catalysts by separately optimizing each elementary step. Overcoming this limitation requires decoupling the binding energies of different intermediates, possibly by constructing active sites with a three-dimensional character that can bind different reaction intermediates in different ways or by selectively stabilizing some of the reaction intermediates via some external mechanism while not stabilizing others.[101,282,283]

8.3 TRANSITION-METAL CATALYSTS

Using DFT calculations, electrochemical ammonia synthesis was investigated on a Ru (0001) catalyst surface at room temperature and atmospheric pressure.[284] The energy diagram for the hydrogenation of N_2 via an associative mechanism on a $MoFe_6S_9$ complex and a Ru (0001) surface. The $MoFe_6S_9$ complex was used to model the active site of FeMo cofactor in the enzyme nitrogenase. It was found that the most stable intermediates on both catalysts were very similar, and there was a relatively large energy barrier for the formation of the N_2H intermediate. As biological N_2 fixation can occur at room temperature, it was proposed that the chemical potential of protons and electrons supplied by the enzyme is higher than that of hydrogen in H_2.[284] Similarly, providing extra energy to the reaction through the electrochemical route results in an increase of the chemical potential of protons and electrons. This further suggests that by applying a negative electric potential, the Ru (0001) surface should be able to produce ammonia at room temperature and atmospheric pressure. However, as protons bind stronger than N_2 molecules to the Ru (0001) surface, hydrogen evolution should compete with ammonia synthesis. N_2 reduction pathways on a stepped Ru surface were explored using DFT calculations.[285] The results indicated that the dissociation of N≡N triple bonds at the early stage of the reaction ($^*N_2 \rightarrow 2^*N$, $^*NNH \rightarrow {}^*N + {}^*NH$, and $^*HNNH \rightarrow 2^*NH$) are kinetically difficult under ambient conditions due to high activation barriers (>1 eV). However, other N – N bond dissociation steps ($^*N_2H_2 \rightarrow {}^*NH_2 + {}^*N$, $^*NH_2NH \rightarrow {}^*NH_2 + {}^*NH$, and $^*NH_2NH_2 \rightarrow 2^*NH_2$) have small activation barriers. As a result, one of the most dominant reaction pathways was expected to be $^*N_2H \rightarrow {}^*HNNH \rightarrow {}^*NH_2NH \rightarrow {}^*NH_2 + {}^*NH \rightarrow 2^*NH_2 \rightarrow {}^*NH_2 + NH_3 \rightarrow 2NH_3$. In addition, in the low overpotential region, hydrogen adsorption was thermodynamically more favorable than N_2 protonation. The Ru stepped surface was covered by $^*H$, which reduces the number of available active sites but also results in an increase in the overpotential for the N_2 reduction reaction.[285] The detailed theoretical evaluation of transition-metal surfaces for electrocatalytic NRR was demonstrated using DFT calculations in combination with a computational standard hydrogen electrode (SHE).[280] The free energy diagrams for the NRR on a range of flat and stepped transition-metal surfaces in contact with an acidic electrolyte (pH = 0) were calculated. The catalytic activity trends were then predicted, assuming that the activation energy barrier scales with the free energy difference for each elementary reaction step.

Analogous to the hydrogen evolution reaction, the combination of adsorbed N_2H_x or NH_x species with H adatoms on the electrode surface is called Tafel-type reactions, and the combination of adsorbed N_2H_x or NH_x species with protons from the solution and electrons from the electrode is called Heyrovsky-type reactions. Heyrovsky-type reactions, in which adsorbed N_2H_x or NH_x species react directly with protons from the solution and electrons from the electrode, were considered for associative and dissociative mechanisms. However, Tafel-type reactions, in which protons from the solution are first reduced on the surface, and then the H adatoms react with the adsorbed N_2H_x or NH_x species, were not considered due to the high activation barriers (≥1 eV) for the Tafel-type reactions ($^*H + {}^*NH_x \rightarrow {}^*NH_x + 1$) for most transition-metal surfaces at room temperature. Volcano-shaped curves were obtained when the minimum theoretical potential for electrochemical ammonia synthesis was plotted as a function of the chemisorption energy of N adatoms on selected transition-metal surfaces (Figure 8.1c). Transition-metal surfaces residing on top of the volcano plot, including Mo, Fe, Rh, and Ru, are estimated to provide the highest catalytic activity for ammonia formation. However, at negative applied bias, the catalyst surfaces will be covered with protons, and hydrogen evolution will be kinetically more favorable than NRR. For those late transition-metal surfaces on the right side of the volcano plots, more negative potentials are required to accomplish the N_2 reduction reaction. In addition, these surfaces are prone to be covered with H adatoms rather than N adatoms at the negative bias, and they are limited by the activation of reactants. Interestingly, calculations indicate early transition-metal surfaces such as Sc, Y, Ti, and Zr, on the left side of the volcano plots, are expected to bind N adatoms more strongly than H adatoms; therefore, significantly higher Faradaic efficiencies toward ammonia production compared with hydrogen evolution are expected if a bias of −1 to −1.5 V vs. SHE is applied on those metal electrodes. In addition, flat surfaces of the early transition metals are predicted to be catalytically more active than the stepped surfaces.

The overpotential requirements in low-temperature electrocatalytic NRR via an associative mechanism using linear scaling relations of intermediates' adsorption energies were calculated using DFT. It was found that the linear scaling between the adsorption energies of two key intermediates in NRR, $^*N_2H$ and $^*NH_2$, require overpotentials of at least −0.5 V to drive the electrochemical reduction of N_2 to ammonia. The theoretical limiting potentials for N_2 reduction and hydrogen evolution as a function of the *N binding energy on selected transition-metal surfaces were compared. The N_2 reduction reaction requires consistently more negative potentials (i.e., larger overpotentials) than hydrogen evolution. The smallest difference in the required overpotentials for N_2 reduction and hydrogen evolution is 0.5 V that is observed for Re(111) (Figure 8.2). These findings from theoretical calculations may explain the lack of high ammonia yield rates on transition metals. To design electrocatalysts with high activity and selectivity for N_2 reduction, the scaling relation has to be broken by selectively stabilizing $^*N_2H$ or destabilizing $^*NH_2$. The electrolyte design should be pursued in addition to engineering catalysts to suppress hydrogen evolution, which is the primary side reaction in electrochemical NRR. For example, using ionic liquids help to increase N_2 solubility and promote NRR at low overpotentials.[286] In addition, controlling the number of protons accessing the catalyst surface may contribute to increased ammonia selectivity.[287] To understand the performance of Pd–Ag nanoparticles toward NRR, the free energy diagram of some key reaction intermediates (i.e., H vacancy ($^*H_{vac}$) and *NNH) is examined using DFT calculations. For pristine Pd, it is known that the α-phase of

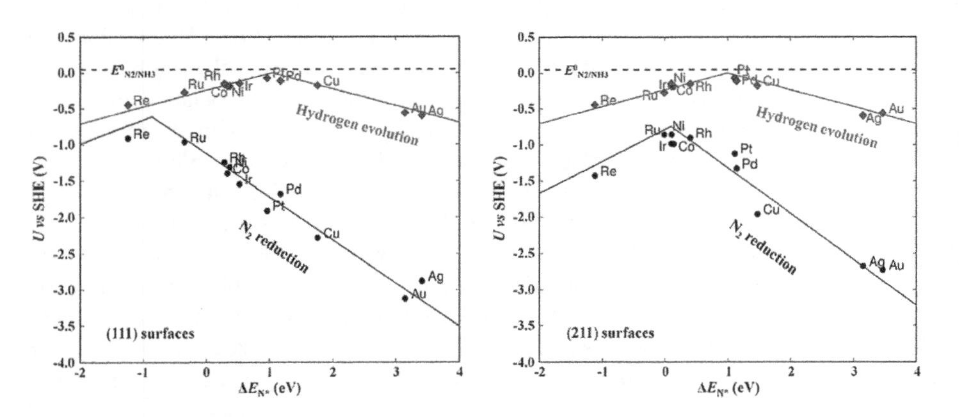

FIGURE 8.2 Comparison of the theoretical limiting potentials required for N_2 reduction and hydrogen evolution for (111) and (211) surfaces. The *N binding energy was chosen as a descriptor for both N_2 reduction and hydrogen evolution because of the N and H adsorption energies scale. Reprinted with permission from Ref.[97]

PdH is the most thermodynamically stable phase under electrochemical operating conditions.[133] The model used for DFT calculations was a (211) surface of PdH where 2/3 of the surface is covered by a monolayer of *H. The formation of H vacancies is required for N_2 to be adsorbed onto active edge sites of the Pd (211) surface. The thermodynamic energy barrier (ΔG) for $^*H_{vac}$ formation is 0.12 eV, followed by an endothermic reaction step of $^*N_2$ hydrogenation with a ΔG value of 1.24 eV. This step is the potential-determining step (PDS). $^*N_2$ hydrogenation is assisted by the Grotthuss-like proton-hopping mechanism present in a hydrated environment.[133] The energy difference for Ag replacement at different sites in the structure is close in energy. Therefore, seven various sites on the PdH (211) surface are examined to study the binding strength of adsorbates after replacing Pd atoms with Ag atoms. When Pd atoms closest to the adsorption sites (sites 1 and 2) are replaced by Ag, the binding energy of *NNH is weakened by 0.01 and 0.36 eV, respectively (Figures 8.3a and 8.3b). In addition, when Pd atoms far from the

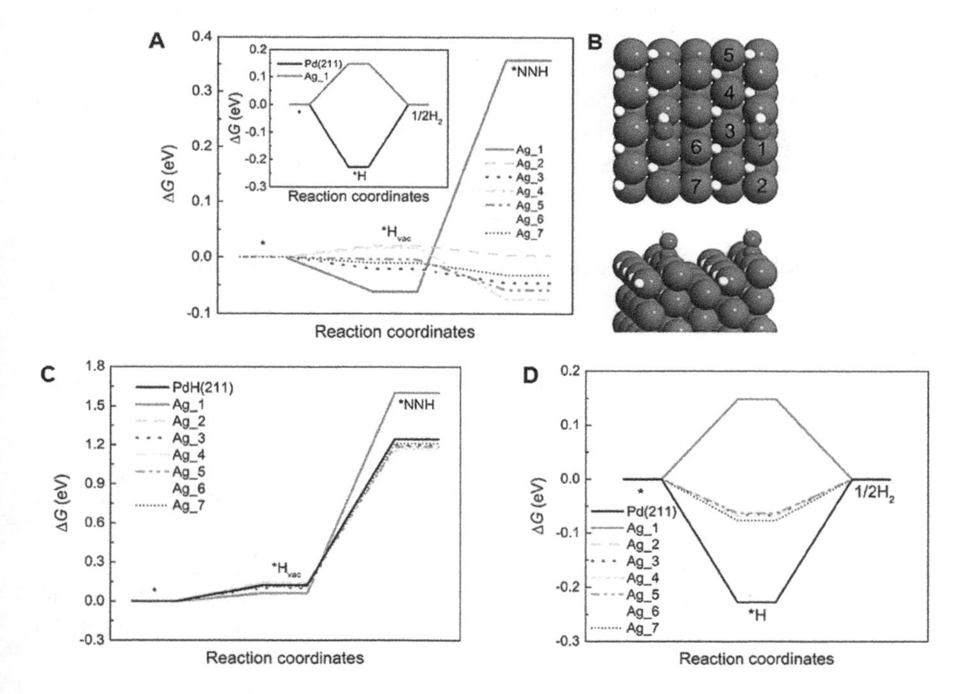

FIGURE 8.3 (a) Calculated free energy diagram for H vacancy formation ($^*H_{vac}$) followed by $^*N_2$ reduction to *NNH on the PdH (211) surface with Ag replacement at different sites. All free energies of $^*H_{vac}$ and *NNH on the PdH (211) surface are used as the energy reference. The inset figure shows the *H binding energy is weakened on the catalyst surface after replacing the Pd atom by Ag atom at site 1. (b) The top and side views of the adsorbed *NNH on PdH (211) surface. Ag_n denotes the different sites when Pd (n) is replaced by Ag. (c) Calculated free energy diagram for the H vacancy formation ($^*H_{vac}$) followed by the $^*N_2$ reduction to *NNH on the PdH (211) surface with Ag replacement at different sites. All the free energies of $^*H_{vac}$ and *NNH on the PdH (211) surface is shown in their absolute value. (d) Calculated free energy diagram for the hydrogen evolution reaction on the PdH (211) surface with Ag replacement at different sites. Reprinted with permission from Ref.[176]

adsorption sites (sites 3–7) are replaced with Ag, the binding energies of *NNH for various adsorption sites are enhanced by ~0.05 eV. While replacing Pd by Ag results in an increase in the thermodynamic energy barrier for $^*N_2$ reduction to *NNH, it weakens the *H binding energy of Pd, raises the ΔG from −0.23 to 0.15 eV, thereby decreasing HER activity on the catalyst surface (Figures 8.3c and 8.3d). Without the presence of Ag atoms, protons would cover most active sites on the Pd nanoparticles. The presence of Ag in bimetallic Pd–Ag nanoparticles suppresses the HER and provides more intrinsically active sites for NRR.

8.4 EX SITU CHARACTERIZATIONS OF CATALYSTS

Electrocatalytic measurements commonly utilize ex situ characterizations, including electron microscopy (e.g., scanning or transmission), X-ray spectroscopy (e.g., photoelectron spectroscopy), and various electrochemical methods (e.g., cyclic voltammetry) to characterize catalysts before and after the reaction. These measurements provide information on catalyst composition, structure, selectivity, and activity toward specific products. However, these techniques cannot provide a comprehensive picture of the electrocatalytic phenomena (e.g., active sites of catalysts or structural changes during the electrochemical reaction) as the catalyst might change under the reaction conditions.[288–290] This informs ex situ and operando characterization techniques to complement ex situ measurements.

8.5 OPERANDO SURFACE-ENHANCED RAMAN SPECTROSCOPY (SERS) MEASUREMENTS

In situ and *operando* surface-enhanced spectroscopy is a technique well suited to probe electrochemical reactions at the electrode/electrolyte interface. Several studies have shown great success in studying electrocatalytic water splitting and CO_2 reduction reactions.[291–293] Surface-enhanced infrared absorption spectroscopy (SEIRAS) was recently used to determine electrocatalytic NRR mechanisms on Au and Pt surfaces.[294] *Operando* surface-enhanced Raman spectroscopy (SERS) allows for the detection of intermediate species even in low abundance and is used for the determination of NRR mechanisms using trimetallic Au–Ag–Pd nanostructures.

Spectroelectrochemistry is a powerful technique that combines spectroscopy and electrochemistry. In this method, a potentiostat is used to apply potential on a SERS active substrate that is then monitored by changes in a spectrum. The Operando SERS cell consisted of machined Teflon parts and fittings outfitted with an Au counter electrode, an Ag/AgCl reference electrode, an Au working electrode with an area of 0.01 cm^2. 1 μL of prepared nanoparticles was drop casted onto Au and left to dry under ambient conditions for 2 h. Before each test, the electrolyte solution is purged with N_2 or Ar for 30 min. Cyclic voltammetry (CV) tests are carried out at the potential range of 0.97 to −0.63 V vs. RHE with the scan rate of 2.5 mV s^{-1} in N_2- or Ar-saturated solution.

To probe the possible reaction mechanisms and track the intermediate species in electrocatalytic NRR, *operando* SERS spectra are collected during the CV tests on the SERS active substrate, which is comprised of Au–Ag–Pd nanoparticles deposited on the Au thin film working electrode. The spectroelectrochemical setup and CV curves are shown in Figure 8.4. The SERS spectra in N_2- and Ar-saturated $LiClO_4$ (aq.) solution contain the vibrational band located at 932 cm^{-1}, which is attributed to the stretching mode of the ClO_4^- anion (Figure 8.5a and 8.5b). In addition, the wide vibrational band centered at 3415 cm^{-1} corresponds to O-H stretching. As the potential is swept to negative values in the reductive pathway (0.97 V to −0.23 V vs. RHE), the faint evolution of three vibrational modes at 1094, 1393, 1601 cm^{-1} is observed, reaching their highest intensity at −0.23 V; these can be classified as N–N stretching, H–N–H bending, and N–H wagging, and these suggest the formation of N_2H_4 as an intermediate species (Figures 8.5a, 8.5c, and 8.5d).[142,294] Two more peaks centered at 2937 cm^{-1} and 3605 cm^{-1} are evolved at −0.23 V, corresponding to the N–H asymmetric and symmetric stretching modes, respectively. The evolution of these two peaks strongly supports the formation of NH_4^+ during the reductive potential sweep.[161,294,295] By moving toward more negative potentials during the reductive CV scan (−0.23 v to −0.63 V), the intensity of peaks at 1094, 1393, 1601 cm^{-1} (intermediates) decreases while the peak at 2937 cm^{-1} and the shoulder at 3605 cm^{-1} (NH_4^+) reach their highest intensities. This strongly suggests the formation of NH_3 from the intermediate species ($N_2 \rightarrow N_2H_4 \rightarrow NH_3$) (Figures 8.5a, 8.5c, and 8.5d). The intensity of the peaks ascribed to intermediates and ammonium decreases during the oxidative pathway due to the oxidation of the N-containing species. The SERS spectra of the standard ammonia and hydrazine solutions in 0.5 M $LiClO_4$ (aq.) confirm the band assignment of various vibrational modes in NRR *operando* SERS measurements (Figure 8.6). Furthermore, *operando* SERS measurements in Ar-saturated electrolyte with the electrocatalyst and N_2-saturated electrolyte without the electrocatalyst reveal no pronounced peaks, implying the peaks observed in

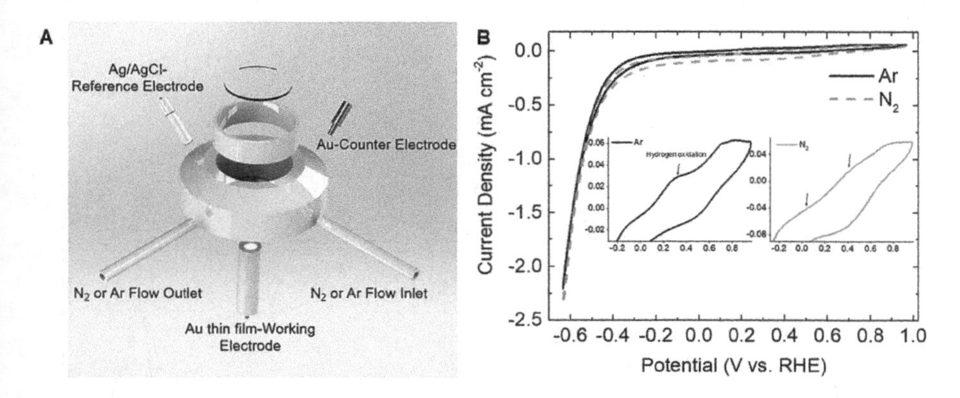

FIGURE 8.4 (a) Schematic of the spectroelectrochemical setup, including Au working and counter electrodes and Ag/AgCl reference electrodes for operando SERS measurements. (b) CV curves of Au–Ag–Pd-850 in Ar- and N_2-saturated 0.5 M $LiClO_4$ (aq.) solution at the scan rate of 2.5 mV s^{-1}. During the oxidation segment in Ar-saturated electrolyte, a strong peak centered on 0.3 V corresponds to the oxidation of hydrogen. This peak with the lower intensity and slight positive potential shift is also observed in N_2-saturated electrolyte. In addition, a new peak at around 0.05 V is observed in N_2-saturated electrolyte, which might be attributed to the oxidation of N-containing species. Reprinted with permission from Ref.[142]

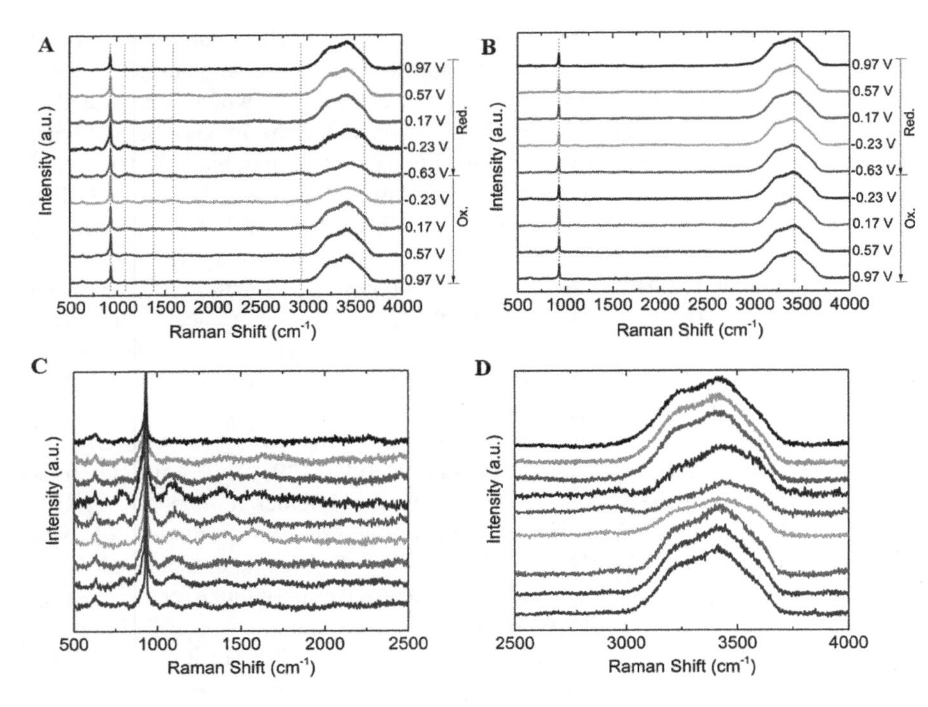

FIGURE 8.5 (a) Operando SERS using Au–Ag–Pd-850 nanoparticles in N_2-saturated 0.5 M $LiClO_4$ (aq.) solution at the scan rate of 2.5 mV s^{-1} with 532 nm laser. (b) Operando SERS using Au–Ag–Pd-850 nanoparticles in Ar-saturated 0.5 M $LiClO_4$ (aq.) solution at the scan rate of 2.5 mV s^{-1} with 532 nm laser. (c and d) Low and high Raman shift of SERS spectra in N_2-saturated electrolyte. Reprinted with permission from Ref.[142]

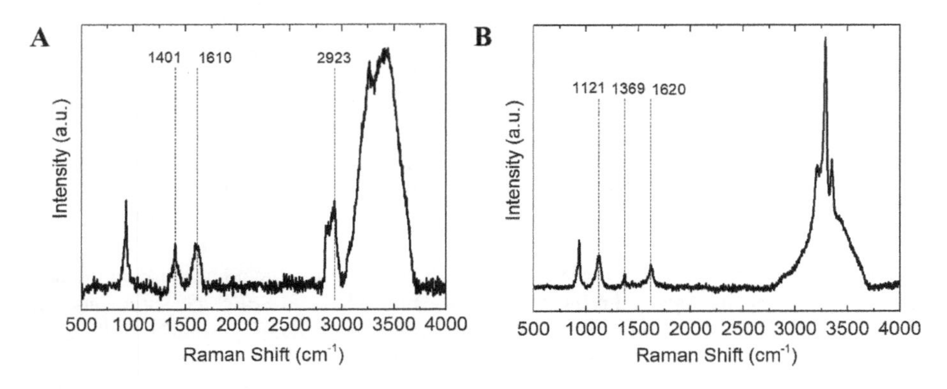

FIGURE 8.6 (a) SERS spectrum of the standard ammonia solution (~30 μM NH_4^+) in 0.5 M $LiClO_4$ (aq.) solution. (b) SERS spectrum of the standard hydrazine solution (~60 μM N_2H_4 (aq.)) in 0.5 M $LiClO_4$ (aq.) solution. Reprinted with permission from Ref.[142]

N_2-saturated electrolyte with the electrocatalyst are solely related to the formation of N-containing species at the presence of the electrocatalyst (Figure 8.5b). In another study, Pd–Ag nanoparticles are used for *operando* SERS investigation during electrochemical ammonia synthesis.[176] During the positive scan (oxidation segment) of the CV test in the N_2-saturated electrolyte, a pronounced peak at ~0.6 V is observed, which is attributed to the oxidation of NH_3 (aq.) produced during NRR (Figure 8.7). Similar to the NRR mechanism on Au–Ag–Pd nanostructures,[142] during the reductive pathway (0.97 V to −0.23 V vs. RHE), three small vibrational peaks at 1101, 1394, and 1613 cm^{-1} are observed, reaching their highest intensity at −0.23 V (Figures 8.8a and 8.8c).[176,294] A strong peak centered at 2936 cm^{-1} appeared as a more negative potential (i.e., −0.63 V) is applied. This peak corresponds to the N-H asymmetric stretching mode. The evolution of this peak supports the formation of NH_4^+ during the reductive potential sweep.[161,294,295] The intensity of the peak for N-H stretching mode using Pd–Ag nanoparticles is significantly higher than that of Au–Ag–Pd nanoparticles, consistent with higher ammonia yield rates on Pd–Ag nanoparticles at lower Faradaic efficiency.[176] The similar NRR mechanism is observed on Pd–Ag nanoparticles, where the NRR proceeds through an associative mechanism with N_2H_4 as an intermediate species (Figures 8.8a, 8.8c, and 8.8d). This also signifies the fact that no detectable amount of hydrazine (N_2H_4) is observed as a by-product in electrochemical NRR studies using Pd-based nanocatalysts.[133,187,191] *Operando* SERS measurements in Ar-saturated electrolyte with the electrocatalyst show no pronounced peaks, implying the peaks observed in the N_2-saturated electrolyte with the electrocatalyst are solely related to the formation of N-containing species at the presence of the Pd–Ag electrocatalyst (Figure 8.8b). Overall, *operando* SERS measurements help understand the reaction mechanism to design more efficient catalysts for electrochemical energy conversion systems.

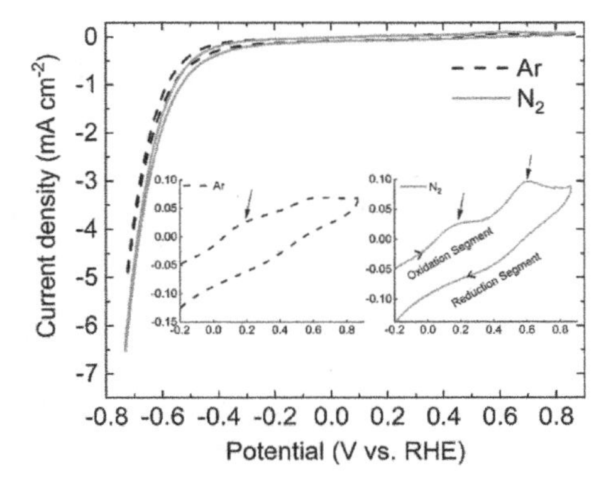

FIGURE 8.7 CV curves of Pd–Ag nanoparticles in Ar- and N_2-saturated 0.5 M $LiClO_4$ (aq.) solution at the scan rate of 2.5 mV s^{-1}. During the oxidation segment in Ar-saturated electrolyte, a strong peak centered on 0.2 V corresponds to the oxidation of hydrogen. This peak with the lower intensity is also observed in N_2-saturated electrolyte. In addition, a new peak at around 0.6 V is observed in N_2-saturated electrolyte, which is attributed to the oxidation of NH_3. Reprinted with permission from Ref.[176]

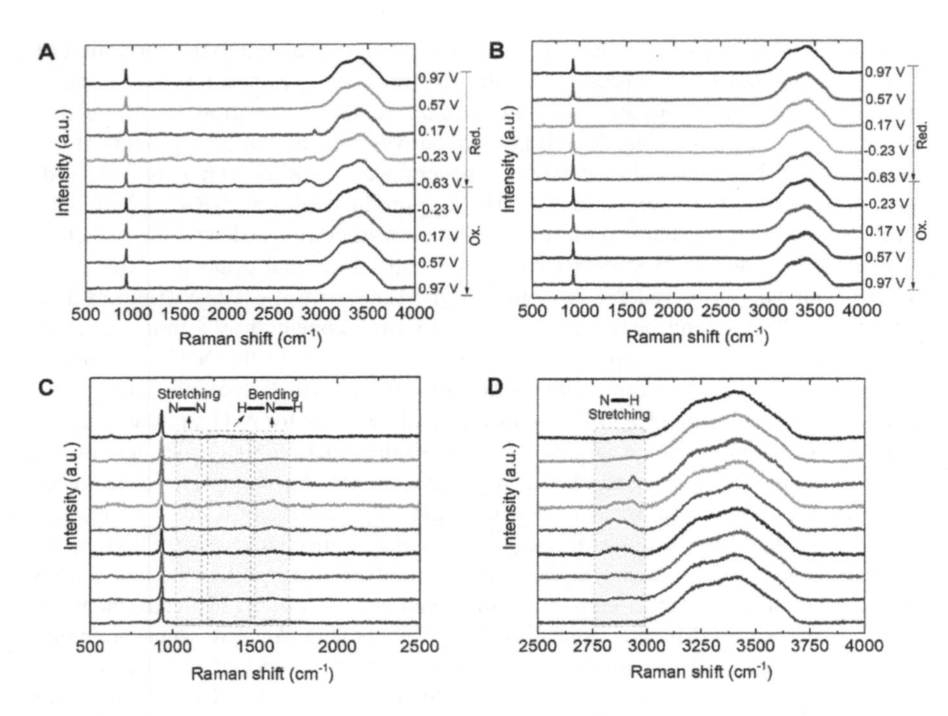

FIGURE 8.8 (a) Operando SERS using Pd–Ag nanoparticles in N_2-saturated 0.5 M $LiClO_4$ (aq.) solution at the scan rate of 2.5 mV s^{-1} with 532 nm laser. (b) Operando SERS using Pd–Ag nanoparticles in Ar-saturated 0.5 M $LiClO_4$ (aq.) solution at the scan rate of 2.5 mV s^{-1} with 532 nm laser. (c and d) Low and high Raman shift of SERS spectra in N_2-saturated electrolyte. Reprinted with permission from Ref.[176]

8.6 OTHER IN SITU SPECTROSCOPIC TECHNIQUES

Probe-based techniques provide spatial resolution to investigate heterogeneous surfaces during electrocatalytic reactions.[296] These techniques provide insights into the local potential in an electric field gradient or probe differences in current density or rate of product generation to identify the active site or the role of defects. X-ray photoelectron spectroscopy (XPS) is a technique that is sensitive to both the chemical environment and electrostatic potential. Core-level photoelectrons are elementally specific and can often distinguish the oxidation state and the surrounding ligand environment. Ambient pressure XPS was used to study a CeO_2_δ solid oxide electrochemical cell driving the CO_2 reduction reaction. XPS was also able to probe the potential profile across a CeO_2_δ catalyst in contact with an Au current collector to identify the corresponding surface speciation and Ce oxidation state.[297] Carbonate and Ce^{3+} accumulated during CO_2 electrolysis to generate CO over a 400-μm active region on the CeO_2_δ.[298] Contact-based potential-sensing electrochemical atomic force microscopy (PS-EC-AFM) is used to probe the potential directly at a conductive tip. This technique provides information regarding local morphology and directly measures the surface electrochemical potential in heterogeneous electrochemical systems. PS-EC-AFM has recently been used to explore the potential- and thickness-dependent electronic properties of cobalt (oxy)hydroxide phosphate (CoPi) on illuminated hematite (α-Fe_2O_3) photoelectrodes.[299]

With sufficient resolution on the atomic scale, probe-based techniques are employed to identify active sites, such as edges or other heterogeneities. Scanning tunneling microscopy (STM) can measure surface reactivity (convoluted with morphology) by the tunneling current. In this technique, highly active sites result in electrochemical noise due to changes in electrolyte composition and local ad/desorption processes (Figure 8.9). This approach has been used to probe the active sites during hydrogen evolution, suggesting that for Pt in acid, step edges are much more active than (111) terraces.[300] In contrast, for metals supported on Au (111), Pt terrace sites are active,

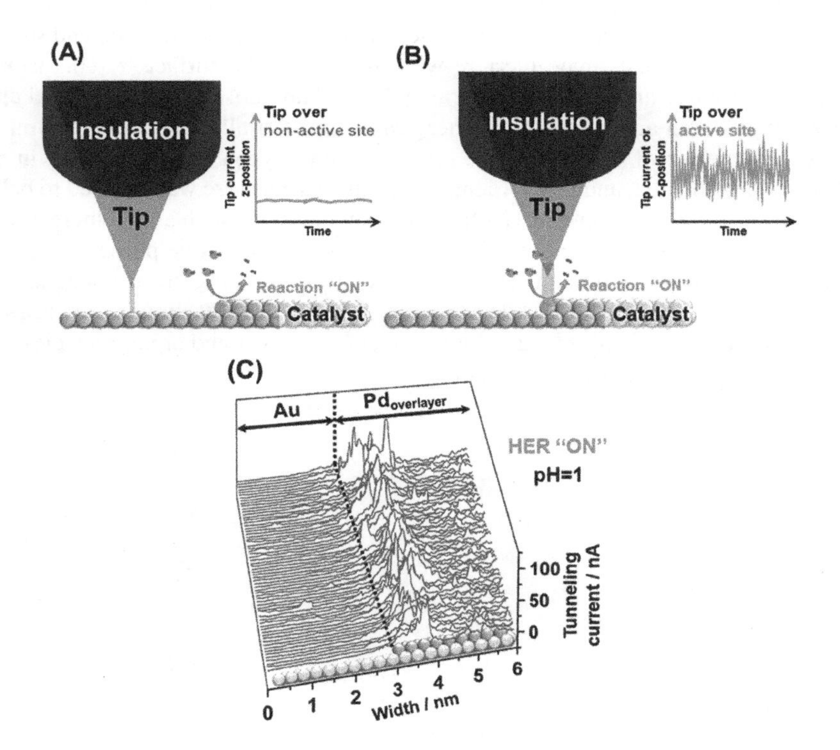

FIGURE 8.9 (a and b) Schematic description of the concept of identifying active electrocatalytic sites. STM can identify active sites by electrochemical noise in the tunneling current. The level of the tunneling current noise should be increased when the STM-tip is located over surface sites with high electrocatalytic activity. Local changes in the environment change the tunneling barrier in STM. Increased tunneling current noise is likely when the tip is over a more active step edge, compared with terrace sites. (c) The boundary between a Pd overlayer on Au and Au surfaces under HER conditions in 0.1 M H_2SO_4. Pd atoms at the boundary are more active than atoms at the center of Pd islands (STM constant height mode). Reprinted with permission from Ref.[301]

whereas Pd atoms at the boundary are more active than those at the center of Pd islands (Figure 8.9).[301]

Scanning electrochemical cell microscopy (SECM) is utilized to demonstrate that the increased catalytic activity of Au grain boundaries for CO_2 and CO reduction is on a length scale (0.5–4 mm), consistent with its dislocation-induced strain field.[302] Dislocation-induced strain may change reaction intermediate binding energy and/or create high step densities with greater activity than terrace sites. Small probes can also be polarized to locally detect products (e.g., evolved H_2 via its oxidation), measuring local activity by the current passed between the probe and the sample. This technique has been employed to study size-dependent nanoparticle activity for HER by an electrochemical STM and an SECM tip.[303,304]

In situ X-ray absorption spectroscopy (XAS) can probe the oxidation state and coordination environment of metal atoms at or near reaction conditions, allowing for more accurate structure-selectivity relationships than ex situ methods. This technique's main limitation comes in surface sensitivity, as XAS probes bulk and surface atoms, and bulk signals may interfere with those from the surface.[305] This lack of surface sensitivity can be mitigated if catalysts used are those with small metal clusters or single-atom active sites, as in these cases, most or all metal atoms are on the surface. The effect of Pt particle size (30–90 Å) on oxygen reduction using in situ XAS was investigated, and it was found that as the particle size was reduced to below 50 Å, the adsorption strength of H, OH, and C_1 compounds such as CO increased.[306]

In situ XAS was also used to identify oxynitride as the active phase for electrochemical nitrogen reduction on vanadium nitride by correlating the disappearance of the oxynitride band with catalyst deactivation.[158] X-ray based techniques are not sensitive to most species in the electric double layer (e.g., water and organic species) due to the discrimination against lighter atoms.

9 Performance Targets and Opportunities for Green Ammonia Synthesis

9.1 INTRODUCTION

Recently, there has been increased importance attached to the development of renewable energy sources to serve as a way to diversify from traditional fossil fuels for sustainable economic growth and for combating the environmental challenges associated with greenhouse gases. Investments are pouring into renewable solutions from various entities across the financial spectrum. However, most research, development, and funding is geared toward energy production and doesn't address energy storage. Currently there is no cost-effective, sustainable method of storing surplus electrical energy for future use on an industrial level other than traditional petrochemical fuels. Extracting cheap electricity in the US, converting it to liquid fuel, and shipping it to net energy importers presents an opportunity to profit on the arbitrage in electrical rates after factoring in all production, construction, and logistical expenses. Concerning its energy storage ability, ammonia is a valuable chemical. Still, it can only be produced in a centralized way with the current manufacturing process known as Haber–Bosch, which is capital- and energy-intensive and is responsible for more than 1% of global CO_2 emissions annually, and consumes >1% of the worldwide energy supply. Micro ammonia production system (MAPS) allows for decentralized and on-site ammonia production with the application for long-term energy storage at the rate of \$0.15–0.2 to sell 1 kWh of electricity with 30–35% gross margin. It is important to note that research and development are ongoing in this field, and it is expected to be possible to lower the cost to sell one kWh of electricity in the foreseeable future.

In the energy market, MAPS can help energy/utility companies to store surplus electricity in fuel (ammonia) using ammonia's chemical properties. The restriction preventing electricity producers from doing so is the total lack of adequate technology to store intermittent renewable energy sources. This technology can complement renewable energy and other electricity producers alike to store surplus generated energy. In addition, ammonia as a carbon-neutral source of energy can be easily transported and converted back to electricity at the point of use, creating an incredibly profitable opportunity to take advantage of the arbitrage between net energy exporters with low-cost electricity and net energy importers with higher electricity rates. This is because ammonia liquefies at −33 °C and stores at 10 bar, which is compared to 250 bar for LNG and 700 bar for liquid hydrogen. This directly correlates to the storage and transmission CAPEX in which ammonia has 4× cheaper

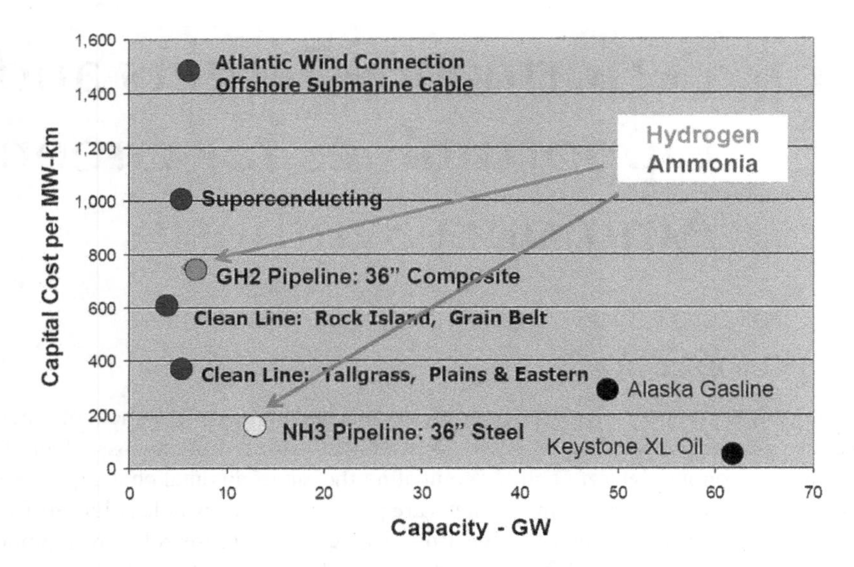

FIGURE 9.1 Comparison of transmission CAPEX of ammonia with various other fuels, including hydrogen and natural gas. The transmission CAPEX for ammonia is quarter that of hydrogen and is also >$100 per MW-km cheaper than natural gas. Reprinted with permission from Ref.[44]

transmission CAPEX compared to H_2 (200 vs. 800 $ per MW-km) and >$100 per MW-km cheaper than LNG (Figure 9.1). Furthermore, with the adaptation of MAPS technology to a system mirroring that of current LNG logistical networks, MAPS can be easily integrated into existing electrical networks with only additional amendments being installment of storage tanks of the liquid ammonia and a retrofitted gas turbine or solid oxide fuel cell (SOFC) adaptor to connect the stored ammonia to the electrical grid of the importing nation.

9.2 MARKET NEED

In recent decades there has been pronounced political, economic, and social importance attached to environmental sustainability, and the repercussions are also extended to the energy sector. These sentiments have translated into policies and memoranda throughout developed and emerging markets alike. Net energy importing regions/nations are mandating increasingly significant portions of their energy consumption to be sourced from renewables to wean themselves of the costly, polluted petroleum-based fuel sources they have been reliant on for the past century. Markets such as Northern Europe, East Asia, Western Africa, and the US States of Hawaii and Alaska, in addition to the US unincorporated territory of Puerto Rico and mainland states such as California, Connecticut, Maine, Massachusetts, New Hampshire, Rhode Island, and Vermont with the current average electricity price of $0.2 per kWh or more can benefit from emerging renewable energy storage technologies for fuel generation.

By sourcing renewable energy in MAPS, one will be able to produce storable energy at remarkably lower prices than what is currently charged in its target importing nations, while simultaneously qualifying for government-sponsored initiatives to encourage renewable energy usage such as but not limited to accelerated amortization, tax breaks, Feed-in-Tariff eligibility, and extremely accommodating loan structures. This enables us to sell its storable energy at a significant markup. At the same time, the importers can purchase the desired, renewable energy sourced power at a rate cheaper than before. This will also enable developed nations concerned about their carbon emission to continue to meet their energy demands, along with developing countries looking to fuel economic growth in an environmentally-friendly, cost-effective, and sustainable manner with significant ease.

Storing renewable energy in the form of fuel could provide energy security at a lucrative profit, all the while offering prices below current real values and sourced with renewable energy for several markets demanding a constant supply of electricity in the stable, profitable markets of Japan, Northern, Southern, and Western Europe, the US state of Hawaii, and the Caribbean Islands, most notably Puerto Rico. We could also provide energy at below-market rates to sustain growth in developing nations in North Africa, East Africa, East Asia, and ASEAN member states.

Each region has its own appropriate opportunity costs associated with doing business in it regarding financing, pricing, and bureaucracy. In the European Union and Eurozone nations, the government uses a Feed-in-Tariff (FiT) to subsidize and incentivize the use of renewable energy. For instance, Germany offers a fixed rate of 1% to fund construction costs of supporting infrastructure with systems similar to MAPS. Furthermore, the FiT model is guaranteed with a 15–20 year contract with the rate per kilowatt-hour remaining constant throughout the contract's life. This is important because as MAPS advances and efficiency increases, the profit spread will increase, assuming inflation does not surpass normal levels. The tradeoff is that rates would have a smaller markup than those of Puerto Rico and Hawaii.

The Japanese model follows a similar FiTs in terms of rate lock-ins and guaranteed contract lengths. However, the Japanese government does not have any incentive programs to assist the funding of such operations, and due to the increased need for renewable energy post-2011 Fukushima incident, the rate on FiT's encompasses an increasingly small profit margin, which can be expected to decrease further.

The markets in the State of Hawaii and the unincorporated territory of Puerto Rico represent much more profitable, stable, willing, and more comfortable markets. Hawaii currently has a combined annual need of >40 billion kWh per year for household and industry consumption, with 85% being supplied by costly, polluted petroleum-based fuel sources, all of which is imported. Due to the state's reliance on international fuel imports, electricity costs are 2.5–3 times more expensive than the continental US, with rates ranging from $0.25 to $0.32 per kWh. These costs and environmental factors combined with the state's commitment to be 100% renewable energy-dependent by 2045 present an opportunity for new renewable energy storage technologies (e.g., MAPS) to supply electricity at a substantial markup to Hawaii while still being able to offer the electricity at a noticeable discount when compared to current prices. Furthermore, the state government has announced its willingness to assist in financing structures and tax incentives to encourage the development of such technologies.

The Unincorporated Territory of Puerto Rico also presents a similar opportunity due to its energy prices, demand, and lack of supply. Hurricane Maria hit the state in 2017, and as a result, most of its energy-generating infrastructure was destroyed or rendered obsolete. Prior to the storm and currently, Puerto Rico has been entirely reliant on petroleum-based fuel sources to generate power to meet its 21 billion kWh energy demand, and this has resulted in prices of electricity being more than twice as high as the continental US, with rates ranging from \$0.21 to \$0.22 per kWh. Furthermore, the government has stated its intention to increase its dependence on renewable energy. Due to the lack of energy generation infrastructure, Federal Emergency Management Agency allocated over 12 billion USD to assist in the reconstruction and the local government's efforts to decrease electricity rates while increasing its renewable energy consumption. This presents an opportunity to provide a substantial portion of Puerto Rico's electricity at a considerable markup while offering rates below what the island currently pays. Furthermore, it is vital to highlight the benefit of both these territories being in the US. This signifies the increased ease of doing business, reduced currency exposures, elimination of double taxation, and a standardized court system.

The US wind power market generated \$11.6 billion in revenue in 2018, expected to grow to \$19.9 billion at a compounded annual growth rate (CAGR) of 11.4% in the next five years to 2023. Likewise, the US Solar Power market generated \$7.5 billion in revenue in 2018 and is expected to grow at 13.8% CAGR in the next five years to 2023. In one manner of approximating the market size for energy storage, one considers Georgia Power's "Nights & Weekends" concept whereby users pay \$0.2 per kWh during peak load times (2–7 pm weekdays) and pay \$0.05 per kWh at all other times. Since peak load energy commands a 4× price premium, if we conservatively assume that storing renewable energy for sale at peak hours leads to a 3× price premium, our served addressable market would be equal to the size of the targeted renewable markets (consisting of the value-add of that storage). Natural gas energy production was a \$50 billion industry in 2018, and since storage and combustion infrastructure can be converted to ammonia, this will not be a limiting factor to our total addressable market. Since wind power is particularly intermittent, we believe this makes a good target market, with a value of \$11.6 billion to our customers, if MAPS can increase the value of wind power generated by 3× through storage for peak demand. This is particularly important for our customers in the New England area (CT, ME, MA, NH, RI, VT) with the average electricity price of \$0.21 per kWh and California (\$0.2 per kWh), Alaska (\$0.24 per kWh), and Hawaii (\$0.32 per kWh). Our back-of-the-envelope calculation suggests that if we feed the SOFC with the ammonia fuel generated through MAPS, the electricity cost can be below \$0.20 per kWh. Our total addressable market in the states with LCOE higher than \$0.15 per kWh is \$58.2 billion (based on the total retail sales in our targeted continental states (388,236,056 MWh) and the LCOE MAPS can offer (\$0.15 per kWh)). It is important to note that in these states, the net electricity generation (supply) is 9.2% lower than the total retail sales (demand). So, technologies such as MAPS may also result in 'energy independence' for these states in electricity supply.

Energy industry stakeholders indicated that early-adopters are likely to be players within renewable energy such as wind who already own infrastructure for both renewable energy and natural gas storage & combustion, such as universities, campuses, or local governments. These parties already place a high premium on energy independence in emergencies. They have shown themselves willing to make infrastructural investments required to install as a MAPS system or a traditional natural gas system. The early majority is likely to consist of power authorities who share the tendency to invest in infrastructure but operate within the public electric grid. These players realize enormous cost savings but will want to see successful implementation before large scale investment.

9.3 PERFORMANCE TARGETS FOR GREEN AMMONIA SYNTHESIS

It is essential to compare the input cost in the electrochemical NRR (i.e., electricity) method for ammonia synthesis with the industrial Haber–Bosch ammonia production process (i.e., natural gas). This helps to better understand the challenges and opportunities for the commercialization of green ammonia synthesis through electrochemical NRR. The amount of electricity that is required to produce a ton of ammonia (1,000 kg ammonia) is determined according to Equation 9.1:

$$E_{elec.} = \frac{i \times V \times t \times 2.78 \times 10^{-10}}{NH_3\ yield \times t \times A} \tag{9.1}$$

where i is the average current density during the constant voltage test (i.e., chronoamperometry (CA) (A)), V is the full cell potential in the electrolysis (V), t is the operation time (s), A is the electrode area (cm^2), and 2.78×10^{-10} is the conversion factor from J to MWh. This gives an input of renewable electricity required to make a ton of ammonia in the electrochemical system. Through the rational design of electrode-electrolyte to decrease electricity consumption in the electrolysis system, further input cost reduction could be achieved. In the Haber–Bosch process, approximately 1000 m^3 (35,314.7 ft^3) of natural gas is needed to make one ton of ammonia. Concerning the price volatility of natural gas throughout the year, the input cost of natural gas to make a ton of ammonia is 143 ± 13:8 USD (US natural gas industrial price is taken from US Energy Information Administration (EIA) report).[307] This suggests that green ammonia synthesis in the electrochemical cell will be cost competitive with the Haber–Bosch process if we can decrease the renewable electricity price to less than \$30 per MWh or \$0.03 per kWh and if we can have an electrochemical cell with the energy input <5 MWh Elec. Per ton of NH_3 (Figure 9.2). This is an optimistic but realistic target to achieve these performance metrics, particularly the electricity price from renewable sources such as solar and wind from the current price of about \$0.04–0.05 per kWh.

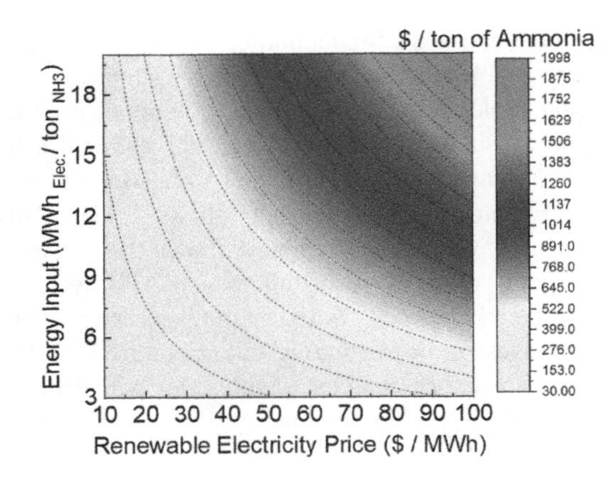

FIGURE 9.2 Input electricity cost for making one ton of green ammonia as a function of renewable electricity price. The input natural gas cost for making a ton of ammonia in the Haber–Bosch process is relatively constant and is about $150, while for green ammonia there is substantial room to decrease the renewable electricity price and therefore be cost competitive with the Haber–Bosch process.

10 Conclusion

Photo-electrosynthesis of ammonia is considered a clean, sustainable, and decentralized approach to the century-old Haber–Bosch process. This approach enables the storage of surplus renewable electricity to a dispatchable and transportable ammonia fuel. Research and development have focused on improving yield and efficiency through the rational design of electrocatalysts, electrolytes (e.g., aqueous and non-aqueous), reaction conditions (e.g., temperature and pressure), and the photo-electrochemical cell. This book discusses the possibility of using ammonia as a multi-purpose fuel to develop and expand clean transportation and energy sectors. Conventional pathways, including thermochemical and biological N_2 fixation, were discussed. The traditional thermochemical process needs to operate at temperatures around 450 °C to improve the reaction kinetics and breaks the N≡N triple bond. It also requires to operate at high pressures (e.g., 150–250 bar) to favor the forward reaction ($N_2 + 3H_2 \leftrightarrow 2NH_3$) toward ammonia formation. Electrocatalytic NRR on transition metal catalysts, including single-atom catalysts, bimetallic, trimetallic, and alloyed nanoparticles, were presented. It was shown that the optimization of the structure, morphology, and composition is critical to enhancing the rate of electro-reduction of N_2 to NH_3. Plasma-driven approaches enable the cleavage of N≡N triple bond and further protonation of N atoms through dissociative mechanisms. However, due to the remarkably high voltage required to initiate the plasma, the energy efficiency of this system is quite low. In addition to the design of electrode-electrolyte, the design of an efficient cell for ammonia production is imperative. The use of gas-phase systems, comprised of a gas diffusion layer and membrane-electrode-assembly, has been demonstrated to be an effective route to decrease ohmic losses and improve energy efficiency. Photocatalytic and photo-electrocatalytic N_2 reduction for ammonia synthesis under ambient conditions using hybrid plasmonic-semiconductor nanoparticles were presented. Combining the plasmonic metal and a visible-light responsive semiconductor increases the photo-generated electrons' concentration upon visible-light illumination, thereby improving the photocatalytic NRR activity. In addition, to overcome the challenge of competing electrons for hydrogen evolution reaction, mediators (e.g., lithium) can be used to decouple the reduction of N_2 from the subsequent hydrogenation to ammonia through the formation of lithium nitride (Li_3N). However, due to the high voltage required for converting lithium ions to lithium and the safety issues related to handling the highly reactive Li metal, new mediators that are low-cost and safe are desirable. Extensive control experiments must be conducted and reported explicitly by each laboratory to avoid false-positive results. Notably, control experiments using $^{15}N_2$ must be performed. It should be noted that $^{15}N_2$ contains considerable amounts of NO_x and NH_x and should be carefully purified with dilute acid (e.g., 1 mM H_2SO_4 (aq.)) and deionized water before use in NRR experiments. DFT calculations and advanced spectroscopic techniques (e.g., SERS) provide insight into the mechanism of N_2 reduction at the electrode-electrolyte interface, leading to the design of more efficient catalysts for NRR. Finally, it was suggested that electrochemical ammonia synthesis using renewable electricity would be cost-competitive with the Haber–Bosch process if the system is operated with the renewable electricity price of less than $30 per MWh or $0.03 per KWh and the electrochemical cell with the energy input <5 MWh Elec. Per ton of NH_3.

References

1. Lewis, L. N.; Lewis, N. Platinum-catalyzed hydrosilylation-colloid formation as the essential step. *Journal of the American Chemical Society* 1986, *108*, 7228–7231.
2. Freund, P. L.; Spiro, M. Colloidal catalysis: the effect of sol size and concentration. *The Journal of Physical Chemistry* 1985, *89*, 1074–1077.
3. Rioux, R.; Song, H.; Hoefelmeyer, J.; Yang, P.; Somorjai, G. High-surface-area catalyst design: synthesis, characterization, and reaction studies of platinum nanoparticles in mesoporous SBA-15 silica. *The Journal of Physical Chemistry B* 2005, *109*, 2192–2202.
4. Ahmadi, T. S.; Wang, Z. L.; Green, T. C.; Henglein, A.; El-Sayed, M. A. Shape-controlled synthesis of colloidal platinum nanoparticles. *Science-New York Then Washington* 1996, 1924–1925.
5. Sun, Y.; Xia, Y. Shape-controlled synthesis of gold and silver nanoparticles. *Science* 2002, *298*, 2176–2179.
6. Burda, C.; Chen, X.; Narayanan, R.; El-Sayed, M. A. Chemistry and properties of nanocrystals of different shapes. *Chemical Reviews* 2005, *105*, 1025–1102.
7. Narayanan, R.; El-Sayed, M. A. Shape-dependent catalytic activity of platinum nanoparticles in colloidal solution. *Nano Letters* 2004, *4*, 1343–1348.
8. Abbet, S.; Ferrari, A. M.; Giordano, L.; Pacchioni, G.; Häkkinen, H.; Landman, U.; Heiz, U. Pd1/MgO (100): a model system in nanocatalysis. *Surface Science* 2002, *514*, 249–255.
9. Narayanan, R.; El-Sayed, M. A. Carbon-supported spherical palladium nanoparticles as potential recyclable catalysts for the Suzuki reaction. *Journal of Catalysis* 2005, *234*, 348–355.
10. Narayanan, R.; Tabor, C.; El-Sayed, M. A. Can the observed changes in the size or shape of a colloidal nanocatalyst reveal the nanocatalysis mechanism type: homogeneous or heterogeneous? *Topics in Catalysis* 2008, *48*, 60.
11. Narayanan, R.; El-Sayed, M. A. FTIR study of the mode of binding of the reactants on the Pd nanoparticle surface during the catalysis of the Suzuki reaction. *The Journal of Physical Chemistry B* 2005, *109*, 4357–4360.
12. Narayanan, R.; El-Sayed, M. A. Changing catalytic activity during colloidal platinum nanocatalysis due to shape changes: electron-transfer reaction. *Journal of the American Chemical Society* 2004, *126*, 7194–7195.
13. Mahmoud, M. A.; Narayanan, R.; El-Sayed, M. A. Enhancing colloidal metallic nanocatalysis: sharp edges and corners for solid nanoparticles and cage effect for hollow ones. *Accounts of Chemical Research* 2013, *46*, 1795–1805.
14. Mahmoud, M.; Saira, F.; El-Sayed, M. Experimental evidence for the nanocage effect in catalysis with hollow nanoparticles. *Nano Letters* 2010, *10*, 3764–3769.
15. Mahmoud, M.; El-Sayed, M. Time dependence and signs of the shift of the surface plasmon resonance frequency in nanocages elucidate the nanocatalysis mechanism in hollow nanoparticles. *Nano Letters* 2011, *11*, 946–953.
16. Mahmoud, M.; El-Sayed, M. Metallic double shell hollow nanocages: the challenges of their synthetic techniques. *Langmuir* 2012, *28*, 4051–4059.
17. Sun, Y.; Mayers, B.; Xia, Y. Metal nanostructures with hollow interiors. *Advanced Materials* 2003, *15*, 641–646.

18. Chen, C.; Kang, Y.; Huo, Z.; Zhu, Z.; Huang, W.; Xin, H. L.; Snyder, J. D.; Li, D.; Herron, J. A.; Mavrikakis, M. Highly crystalline multimetallic nanoframes with three-dimensional electrocatalytic surfaces. *Science* 2014, *343*, 1339–1343.
19. Zeng, J.; Zhang, Q.; Chen, J.; Xia, Y. A comparison study of the catalytic properties of Au-based nanocages, nanoboxes, and nanoparticles. *Nano Letters* 2010, *10*, 30–35.
20. Weng, G.; Mahmoud, M. A.; El-Sayed, M. A. Nanocatalysts can change the number of electrons involved in oxidation–reduction reaction with the nanocages being the most efficient. *The Journal of Physical Chemistry C* 2012, *116*, 24171–24176.
21. Yen, C.; Mahmoud, M.; El-Sayed, M. Photocatalysis in gold nanocage nanoreactors. *The Journal of Physical Chemistry A* 2009, *113*, 4340–4345.
22. Yadav, M.; Akita, T.; Tsumori, N.; Xu, Q. Strong metal–molecular support interaction (SMMSI): amine-functionalized gold nanoparticles encapsulated in silica nanospheres highly active for catalytic decomposition of formic acid. *Journal of Materials Chemistry* 2012, *22*, 12582–12586.
23. Eddaoudi, M.; Kim, J.; Rosi, N.; Vodak, D.; Wachter, J.; O'Keeffe, M.; Yaghi, O. M. Systematic design of pore size and functionality in isoreticular MOFs and their application in methane storage. *Science* 2002, *295*, 469–472.
24. Graeser, M.; Pippel, E.; Greiner, A.; Wendorff, J. H. Polymer Core–Shell Fibers with Metal Nanoparticles as Nanoreactor for Catalysis. *Macromolecules* 2007, *40*, 6032–6039.
25. Park, J. C.; Bang, J. U.; Lee, J.; Ko, C. H.; Song, H. Ni@ SiO_2 yolk-shell nanoreactor catalysts: high temperature stability and recyclability. *Journal of Materials Chemistry* 2010, *20*, 1239–1246.
26. Macdonald, J. E.; Sadan, M. B.; Houben, L.; Popov, I.; Banin, U. Hybrid nanoscale inorganic cages. *Nature Materials* 2010, *9*, 810–815.
27. Rodríguez-Lorenzo, L.; Alvarez-Puebla, R. A.; Pastoriza-Santos, I.; Mazzucco, S.; Stéphan, O.; Kociak, M.; Liz-Marzán, L. M.; García de Abajo, F. J. Zeptomol detection through controlled ultrasensitive surface-enhanced Raman scattering. *Journal of the American Chemical Society* 2009, *131*, 4616–4618.
28. Li, J. F.; Huang, Y. F.; Ding, Y.; Yang, Z. L.; Li, S. B.; Zhou, X. S.; Fan, F. R.; Zhang, W.; Zhou, Z. Y.; Ren, B. Shell-isolated nanoparticle-enhanced Raman spectroscopy. *Nature* 2010, *464*, 392–395.
29. Heck, K. N.; Janesko, B. G.; Scuseria, G. E.; Halas, N. J.; Wong, M. S. Using catalytic and surface-enhanced Raman spectroscopy-active gold nanoshells to understand the role of basicity in glycerol oxidation. *ACS Catalysis* 2013, *3*, 2430–2435.
30. Smil, V. *Enriching the earth: Fritz Haber, Carl Bosch, and the transformation of world food production*. MIT Press; 2004.
31. Ober, J. A. *Mineral commodity summaries 2016*; 2016.
32. Smil, V. Detonator of the population explosion. *Nature* 1999, *400*, 415.
33. Kalyani, D.. Fertilizer manufacturing in the US: US industry market research report. *IBISWorld* (Feb. 2018). https://www.ibisworld.com/industry-trends/market-research-reports/manufacturing/chemical/fertilizer-manufacturing.html. 2018.
34. Mosheim, R. Fertilizer use and price. *U.S. Department of Agriculture* (Feb. 21, 2018). https://www.ers.usda.gov/data-products/fertilizer-use-and-price.aspx. 2018.
35. Lan, R.; Tao, S. Ammonia as a suitable fuel for fuel cells. *Frontiers in Energy Research* 2014, *2*, 35.
36. Giddey, S.; Badwal, S.; Munnings, C.; Dolan, M. Ammonia as a renewable energy transportation media. *ACS Sustainable Chemistry & Engineering* 2017, *5*, 10231–10239.
37. Hua, T.; Ahluwalia, R.; Peng, J.-K.; Kromer, M.; Lasher, S.; McKenney, K.; Law, K.; Sinha, J. Technical assessment of compressed hydrogen storage tank systems for automotive applications. *International Journal of Hydrogen Energy* 2011, *36*, 3037–3049.

38. Strait, R.; Nagvekar, M. Carbon dioxide capture and storage in the nitrogen and syngas industries. *Nitrogen+ Syngas* 2010, *303*, 1–3.
39. MacFarlane, D. R.; Cherepanov, P. V.; Choi, J.; Suryanto, B. H.; Hodgetts, R. Y.; Bakker, J. M.; Vallana, F. M. F.; Simonov, A. N. A Roadmap to the ammonia economy. *Joule* 2020, *4*, 1186–1205.
40. Brown, T. The AmVeh – An ammonia fueled car from south Korea. *Ammonia Energy Association.* https://nh3fuelassociation.org/2013/06/20/the-amveh-an-ammonia-fueled-car-from-south-korea/. 2013.
41. de Vries, N. *Safe and effective application of ammonia as a marine fuel*; 2019.
42. Abbasov, F. Transport & Environment Group. *Roadmap to decarbonising European shipping.* https://www.transportenvironment.org/publications/roadmap-decarbonising-european-shipping. 2018.
43. Nakatsuka, N., Fukui, J., Tainaka, K., Higashino, H. H. J.; Akamatsu, F. *Detailed observation of coal-ammonia cocombustion processes.* https://nh3fuelassociation.org/2017/09/27/detailed-observation-of-coal-ammonia-co-combustion-processes/. 2017.
44. Valera-Medina, A.; Xiao, H.; Owen-Jones, M.; David, W.; Bowen, P. Ammonia for power. *Progress in Energy and Combustion Science* 2018, *69*, 63–102.
45. Lehnert, N.; Dong, H. T.; Harland, J. B.; Hunt, A. P.; White, C. J. Reversing nitrogen fixation. *Nature Reviews Chemistry* 2018, *2*, 278–289.
46. Chen, J. G.; Crooks, R. M.; Seefeldt, L. C.; Bren, K. L.; Bullock, R. M.; Darensbourg, M. Y.; Holland, P. L.; Hoffman, B.; Janik, M. J.; Jones, A. K. Beyond fossil fuel–driven nitrogen transformations. *Science* 2018, *360.*
47. Burgess, B. K.; Lowe, D. J. Mechanism of molybdenum nitrogenase. *Chemical Reviews* 1996, *96*, 2983–3012.
48. Hoffman, B. M.; Lukoyanov, D.; Yang, Z.-Y.; Dean, D. R.; Seefeldt, L. C. Mechanism of nitrogen fixation by nitrogenase: the next stage. *Chemical Reviews* 2014, *114*, 4041–4062.
49. Burford, R. J.; Fryzuk, M. D. Examining the relationship between coordination mode and reactivity of dinitrogen. *Nature Reviews Chemistry* 2017, *1*, 1–13.
50. Erisman, J. W.; Sutton, M. A.; Galloway, J.; Klimont, Z.; Winiwarter, W. How a century of ammonia synthesis changed the world. *Nature Geoscience* 2008, *1*, 636–639.
51. Godfray, H. C. J.; Beddington, J. R.; Crute, I. R.; Haddad, L.; Lawrence, D.; Muir, J. F.; Pretty, J.; Robinson, S.; Thomas, S. M.; Toulmin, C. Food security: the challenge of feeding 9 billion people. *Science* 2010, *327*, 812–818.
52. Foster, S. L.; Bakovic, S. I. P.; Duda, R. D.; Maheshwari, S.; Milton, R. D.; Minteer, S. D.; Janik, M. J.; Renner, J. N.; Greenlee, L. F. Catalysts for nitrogen reduction to ammonia. *Nature Catalysis* 2018, *1*, 490–500.
53. Spatzal, T.; Aksoyoglu, M.; Zhang, L.; Andrade, S. L.; Schleicher, E.; Weber, S.; Rees, D. C.; Einsle, O. Evidence for interstitial carbon in nitrogenase FeMo cofactor. *Science* 2011, *334*, 940–940.
54. Doan, P. E.; Telser, J.; Barney, B. M.; Igarashi, R. Y.; Dean, D. R.; Seefeldt, L. C.; Hoffman, B. M. 57Fe ENDOR spectroscopy and 'electron inventory' analysis of the nitrogenase E4 intermediate suggest the metal-ion core of FeMo-cofactor cycles through only one redox couple. *Journal of the American Chemical Society* 2011, *133*, 17329–17340.
55. Bjornsson, R.; Lima, F. A.; Spatzal, T.; Weyhermüller, T.; Glatzel, P.; Bill, E.; Einsle, O.; Neese, F.; DeBeer, S. Identification of a spin-coupled Mo (III) in the nitrogenase iron–molybdenum cofactor. *Chemical Science* 2014, *5*, 3096–3103.
56. Bjornsson, R.; Neese, F.; Schrock, R. R.; Einsle, O.; DeBeer, S. The discovery of Mo (III) in FeMoco: reuniting enzyme and model chemistry. *JBIC Journal of Biological Inorganic Chemistry* 2015, *20*, 447–460.

57. Varley, J.; Wang, Y.; Chan, K.; Studt, F.; Nørskov, J. Mechanistic insights into nitrogen fixation by nitrogenase enzymes. *Physical Chemistry Chemical Physics* 2015, *17*, 29541–29547.
58. Spatzal, T.; Perez, K. A.; Einsle, O.; Howard, J. B.; Rees, D. C. Ligand binding to the FeMo-cofactor: structures of CO-bound and reactivated nitrogenase. *Science* 2014, *345*, 1620–1623.
59. Rao, L.; Xu, X.; Adamo, C. Theoretical investigation on the role of the central carbon atom and close protein environment on the nitrogen reduction in Mo nitrogenase. *ACS Catalysis* 2016, *6*, 1567–1577.
60. FAO-Food; Nations, A. O. o. t. U. *World fertilizer trends and outlook to 2020*; Food and Agriculture Organization of the United Nations (FAO), Rome, Italy. 2017.
61. Smith, C.; Hill, A. K.; Torrente-Murciano, L. Current and future role of Haber–Bosch ammonia in a carbon-free energy landscape. *Energy & Environmental Science* 2020, *13*, 331–344.
62. Schiffer, Z. J.; Manthiram, K. Electrification and decarbonization of the chemical industry. *Joule* 2017, *1*, 10–14.
63. Vojvodic, A.; Medford, A. J.; Studt, F.; Abild-Pedersen, F.; Khan, T. S.; Bligaard, T.; Nørskov, J. Exploring the limits: a low-pressure, low-temperature Haber–Bosch process. *Chemical Physics Letters* 2014, *598*, 108–112.
64. Spencer, M. On the rate-determining step and the role of potassium in the catalytic synthesis of ammonia. *Catalysis Letters* 1992, *13*, 45–53.
65. Ertl, G.; Prigge, D.; Schloegl, R.; Weiss, M. Surface characterization of ammonia synthesis catalysts. *Journal of Catalysis* 1983, *79*, 359–377.
66. Ozaki, A.; Taylor, H. S.; Boudart, M. Kinetics and mechanism of the ammonia synthesis. *Proceedings of the Royal Society of London. Series A. Mathematical and Physical Sciences* 1960, *258*, 47–62.
67. Brown, D. E.; Edmonds, T.; Joyner, R. W.; McCarroll, J. J.; Tennison, S. R. The genesis and development of the commercial BP doubly promoted catalyst for ammonia synthesis. *Catalysis Letters* 2014, *144*, 545–552.
68. Liu, H. Ammonia synthesis catalyst 100 years: practice, enlightenment and challenge. *Chinese Journal of Catalysis* 2014, *35*, 1619–1640.
69. Hrbek, J. Coadsorption of oxygen and hydrogen on ruthenium (001): blocking and electronic effects of preadsorbed oxygen. *The Journal of Physical Chemistry* 1986, *90*, 6217–6222.
70. Jacobsen, C. J. Boron nitride: a novel support for ruthenium-based ammonia synthesis catalysts. *Journal of Catalysis* 2001, *200*, 1–3.
71. Yang, X.-L.; Zhang, W.-Q.; Xia, C.-G.; Xiong, X.-M.; Mu, X.-Y.; Hu, B. Low temperature ruthenium catalyst for ammonia synthesis supported on $BaCeO_3$ nanocrystals. *Catalysis Communications* 2010, *11*, 867–870.
72. Niwa, Y.; Aika, K.-I. The effect of lanthanide oxides as a support for ruthenium catalysts in ammonia synthesis. *Journal of Catalysis* 1996, *162*, 138–142.
73. Saito, M.; Itoh, M.; Iwamoto, J.; Li, C.-Y.; Machida, K.-I. Synergistic effect of MgO and CeO_2 as a support for ruthenium catalysts in ammonia synthesis. *Catalysis Letters* 2006, *106*, 107–110.
74. Liang, C.; Li, Z.; Qiu, J.; Li, C. Graphitic nanofilaments as novel support of Ru–Ba catalysts for ammonia synthesis. *Journal of Catalysis* 2002, *211*, 278–282.
75. Fishel, C. T.; Davis, R. J.; Garces, J. M. Ammonia synthesis catalyzed by ruthenium supported on basic zeolites. *Journal of Catalysis* 1996, *163*, 148–157.
76. Hara, M.; Kitano, M.; Hosono, H. Ru-loaded $C_{12}A_7$: e–electride as a catalyst for ammonia synthesis. *ACS Catalysis* 2017, *7*, 2313–2324.

77. Bicer, Y.; Dincer, I.; Zamfirescu, C.; Vezina, G.; Raso, F. Comparative life cycle assessment of various ammonia production methods. *Journal of Cleaner Production* 2016, *135*, 1379–1395.
78. Boulamanti, A.; Moya, J. A. Energy efficiency and GHG emissions: prospective scenarios for the chemical and petrochemical industry. *Report 9789279657344, EU Science Hub*. 2017.
79. Dudley, B. BP statistical review of world energy. *BP Statistical Review*, London, UK, accessed Aug. 2018, *6*, 2018.
80. Soloveichik, G. Electrochemical synthesis of ammonia as a potential alternative to the Haber–Bosch process. *Nature Catalysis* 2019, *2*, 377.
81. Zamfirescu, C.; Dincer, I. Using ammonia as a sustainable fuel. *Journal of Power Sources* 2008, *185*, 459–465.
82. Klerke, A.; Christensen, C. H.; Nørskov, J. K.; Vegge, T. Ammonia for hydrogen storage: challenges and opportunities. *Journal of Materials Chemistry* 2008, *18*, 2304–2310.
83. Lan, R.; Irvine, J. T.; Tao, S. Synthesis of ammonia directly from air and water at ambient temperature and pressure. *Scientific Reports* 2013, *3*, 1145.
84. Chen, S.; Perathoner, S.; Ampelli, C.; Mebrahtu, C.; Su, D.; Centi, G. Electrocatalytic synthesis of ammonia at room temperature and atmospheric pressure from water and nitrogen on a carbon-nanotube-based electrocatalyst. *Angewandte Chemie* 2017, *129*, 2743–2747.
85. Chen, G.-F.; Cao, X.; Wu, S.; Zeng, X.; Ding, L.-X.; Zhu, M.; Wang, H. Ammonia electrosynthesis with high selectivity under ambient conditions via a Li+ incorporation strategy. *Journal of the American Chemical Society* 2017, *139*, 9771–9774.
86. Liu, Y.; Su, Y.; Quan, X.; Fan, X.; Chen, S.; Yu, H.; Zhao, H.; Zhang, Y.; Zhao, J. Facile ammonia synthesis from electrocatalytic N_2 reduction under ambient conditions on N-doped porous carbon. *ACS Catalysis* 2018, *8*, 1186–1191.
87. Hu, L.; Khaniya, A.; Wang, J.; Chen, G.; Kaden, W. E.; Feng, X. Ambient electrochemical ammonia synthesis with high selectivity on Fe/Fe oxide catalyst. *ACS Catalysis* 2018, *8*, 9312–9319.
88. Cui, X.; Tang, C.; Zhang, Q. A review of electrocatalytic reduction of dinitrogen to ammonia under ambient conditions. *Advanced Energy Materials* 2018, *8*, 1800369.
89. Kim, K.; Kim, J.-N.; Yoon, H. C.; Han, J.-I. Effect of electrode material on the electrochemical reduction of nitrogen in a molten LiCl–KCl–CsCl system. *International Journal of Hydrogen Energy* 2015, *40*, 5578–5582.
90. Licht, S.; Cui, B.; Wang, B.; Li, F.-F.; Lau, J.; Liu, S. Ammonia synthesis by N_2 and steam electrolysis in molten hydroxide suspensions of nanoscale Fe_2O_3. *Science* 2014, *345*, 637–640.
91. Imamura, K.; Kubota, J. Electrochemical membrane cell for NH_3 synthesis from N_2 and H_2O by electrolysis at 200 to 250° C using a Ru catalyst, hydrogen-permeable Pd membrane and phosphate-based electrolyte. *Sustainable Energy & Fuels* 2018, *2*, 1278–1286.
92. Amar, I. A.; Lan, R.; Petit, C. T.; Tao, S. Electrochemical synthesis of ammonia based on CO_3 Mo_3N catalyst and $LiAlO_2$–(Li, Na, K)$_2$ CO_3 composite electrolyte. *Electrocatalysis* 2015, *6*, 286–294.
93. Amar, I. A.; Lan, R.; Petit, C. T.; Tao, S. Solid-state electrochemical synthesis of ammonia: a review. *Journal of Solid State Electrochemistry* 2011, *15*, 1845.
94. Kyriakou, V.; Garagounis, I.; Vasileiou, E.; Vourros, A.; Stoukides, M. Progress in the electrochemical synthesis of ammonia. *Catalysis Today* 2017, *286*, 2–13.
95. Singh, A. R.; Rohr, B. A.; Statt, M. J.; Schwalbe, J. A.; Cargnello, M.; Nørskov, J. K. Strategies toward selective electrochemical ammonia synthesis. *ACS Catalysis* 2019, *9*, 8316–8324.

96. Singh, A. R.; Rohr, B. A.; Schwalbe, J. A.; Cargnello, M.; Chan, K.; Jaramillo, T. F.; Chorkendorff, I.; Nørskov, J. K. Electrochemical ammonia synthesis: the selectivity challenge. *ACS Catalysis* 2017, *7*, 706–709.
97. Montoya, J. H.; Tsai, C.; Vojvodic, A.; Nørskov, J. K. The challenge of electrochemical ammonia synthesis: a new perspective on the role of nitrogen scaling relations. *ChemSusChem* 2015, *8*, 2180–2186.
98. Giddey, S.; Badwal, S.; Kulkarni, A. Review of electrochemical ammonia production technologies and materials. *International Journal of Hydrogen Energy* 2013, *38*, 14576–14594.
99. Qing, G.; Hamann, T. W. New electrolytic devices produce ammonia with exceptional selectivity. *Joule* 2019, *3*, 634–636.
100. Hawtof, R.; Ghosh, S.; Guarr, E.; Xu, C.; Sankaran, R. M.; Renner, J. N. Catalyst-free, highly selective synthesis of ammonia from nitrogen and water by a plasma electrolytic system. *Science Advances* 2019, *5*, eaat5778.
101. Qing, G.; Ghazfar, R.; Jackowski, S. T.; Habibzadeh, F.; Ashtiani, M. M.; Chen, C.-P.; Smith III, M. R.; Hamann, T. W. Recent advances and challenges of electrocatalytic N_2 reduction to ammonia. *Chemical Reviews* 2020, *120*, 5437–5516.
102. Zhan, C.-G.; Nichols, J. A.; Dixon, D. A. Ionization potential, electron affinity, electronegativity, hardness, and electron excitation energy: molecular properties from density functional theory orbital energies. *The Journal of Physical Chemistry A* 2003, *107*, 4184–4195.
103. Ertl, G. Reactions at surfaces: from atoms to complexity (Nobel Lecture). *Angewandte Chemie International Edition* 2008, *47*, 3524–3535.
104. Shi, R.; Zhang, X.; Waterhouse, G. I.; Zhao, Y.; Zhang, T. The journey toward low temperature, low pressure catalytic nitrogen fixation. *Advanced Energy Materials* 2020, *10*, 2000659.
105. Guo, W.; Zhang, K.; Liang, Z.; Zou, R.; Xu, Q. Electrochemical nitrogen fixation and utilization: theories, advanced catalyst materials and system design. *Chemical Society Reviews* 2019, *48*, 5658–5716.
106. Tsuneto, A.; Kudo, A.; Sakata, T. Efficient electrochemical reduction of N_2 to NH_3 catalyzed by lithium. *Chemistry Letters* 1993, *22*, 851–854.
107. Tsuneto, A.; Kudo, A.; Sakata, T. Lithium-mediated electrochemical reduction of high pressure N_2 to NH_3. *Journal of Electroanalytical Chemistry* 1994, *367*, 183–188.
108. McEnaney, J. M.; Singh, A. R.; Schwalbe, J. A.; Kibsgaard, J.; Lin, J. C.; Cargnello, M.; Jaramillo, T. F.; Nørskov, J. K. Ammonia synthesis from N_2 and H_2O using a lithium cycling electrification strategy at atmospheric pressure. *Energy & Environmental Science* 2017, *10*, 1621–1630.
109. Lazouski, N.; Chung, M.; Williams, K.; Gala, M. L.; Manthiram, K. Non-aqueous gas diffusion electrodes for rapid ammonia synthesis from nitrogen and water-splitting-derived hydrogen. *Nature Catalysis* 2020, *3*, 463–469.
110. Yao, Y.; Zhu, S.; Wang, H.; Li, H.; Shao, M. A Spectroscopic study on the nitrogen electrochemical reduction reaction on gold and platinum surfaces. *Journal of the American Chemical Society* 2018, *140*, 1496–1501.
111. Bao, D.; Zhang, Q.; Meng, F. L.; Zhong, H. X.; Shi, M. M.; Zhang, Y.; Yan, J. M.; Jiang, Q.; Zhang, X. B. Electrochemical reduction of N_2 under ambient conditions for artificial N_2 fixation and renewable energy storage using N_2/NH_3 cycle. *Advanced Materials* 2017, *29*, 1604799.
112. Shi, M. M.; Bao, D.; Wulan, B. R.; Li, Y. H.; Zhang, Y. F.; Yan, J. M.; Jiang, Q. Au subnanoclusters on TiO_2 toward highly efficient and selective electrocatalyst for N_2 conversion to NH_3 at ambient conditions. *Advanced Materials* 2017, *29*, 1606550.

113. Mahmoud, M. A.; O'Neil, D.; El-Sayed, M. A. Hollow and solid metallic nanoparticles in sensing and in nanocatalysis. *Chemistry of Materials* 2013, *26*, 44–58.
114. Nazemi, M.; Panikkanvalappil, S. R.; El-Sayed, M. A. Enhancing the rate of electrochemical nitrogen reduction reaction for ammonia synthesis under ambient conditions using hollow gold nanocages. *Nano Energy* 2018, *49*, 316–323.
115. Bordley, J. A.; El-Sayed, M. A. Enhanced electrocatalytic activity toward the oxygen reduction reaction through alloy formation: platinum–silver alloy nanocages. *The Journal of Physical Chemistry C* 2016, *120*, 14643–14651.
116. Song, Y.; Johnson, D.; Peng, R.; Hensley, D. K.; Bonnesen, P. V.; Liang, L.; Huang, J.; Yang, F.; Zhang, F.; Qiao, R. A physical catalyst for the electrolysis of nitrogen to ammonia. *Science Advances* 2018, *4*, e1700336.
117. Oshikiri, T.; Ueno, K.; Misawa, H. Selective dinitrogen conversion to ammonia using water and visible light through plasmon-induced charge separation. *Angewandte Chemie* 2016, *128*, 4010–4014.
118. Stiles, P. L.; Dieringer, J. A.; Shah, N. C.; Van Duyne, R. P. Surface-enhanced Raman spectroscopy. *Annual Review of Analytical Chemistry* 2008, *1*, 601–626.
119. Kydd, R. A.; Cooney, R. P. Raman spectra of adsorbed ammonia on carbon-overlayered silver electrodes. *Journal of the Chemical Society, Faraday Transactions 1: Physical Chemistry in Condensed Phases* 1983, *79*, 2887–2897.
120. Dong, J.-L.; Li, X.-H.; Zhao, L.-J.; Xiao, H.-S.; Wang, F.; Guo, X.; Zhang, Y.-H. raman observation of the interactions between NH_4+, SO_{42}–, and H_2O in supersaturated (NH_4) $2SO_4$ droplets. *The Journal of Physical Chemistry B* 2007, *111*, 12170–12176.
121. Zhou, F.; Azofra, L. M.; Ali, M.; Kar, M.; Simonov, A. N.; McDonnell-Worth, C.; Sun, C.; Zhang, X.; MacFarlane, D. R. Electro-synthesis of ammonia from nitrogen at ambient temperature and pressure in ionic liquids. *Energy & Environmental Science* 2017, *10*, 2516–2520.
122. Nash, J.; Yang, X.; Anibal, J.; Wang, J.; Yan, Y.; Xu, B. Electrochemical nitrogen reduction reaction on noble metal catalysts in proton and hydroxide exchange membrane electrolyzers. *Journal of The Electrochemical Society* 2017, *164*, F1712–F1716.
123. Nazemi, M.; El-Sayed, M. A. Electrochemical synthesis of ammonia from N_2 and H_2O under ambient conditions using pore-size-controlled hollow gold nanocatalysts with tunable plasmonic properties. *The Journal of Physical Chemistry Letters* 2018, *9*, 5160–5166.
124. Zhang, Z.; Xin, L.; Li, W. Supported gold nanoparticles as anode catalyst for anion-exchange membrane-direct glycerol fuel cell (AEM-DGFC). *International Journal of Hydrogen Energy* 2012, *37*, 9393–9401.
125. Nazemi, M.; El-Sayed, M. A. The role of oxidation of silver in bimetallic gold–silver nanocages on electrocatalytic activity of nitrogen reduction reaction. *The Journal of Physical Chemistry C* 2019, *123*, 11422–11427.
126. Waterhouse, G. I.; Bowmaker, G. A.; Metson, J. B. The thermal decomposition of silver (I, III) oxide: a combined XRD, FT-IR and Raman spectroscopic study. *Physical Chemistry Chemical Physics* 2001, *3*, 3838–3845.
127. Polavarapu, L.; Zanaga, D.; Altantzis, T.; Rodal-Cedeira, S.; Pastoriza-Santos, I.; Pérez-Juste, J.; Bals, S.; Liz-Marzán, L. M. Galvanic replacement coupled to seeded growth as a route for shape-controlled synthesis of plasmonic nanorattles. *Journal of the American Chemical Society* 2016, *138*, 11453–11456.
128. Goris, B.; Polavarapu, L.; Bals, S.; Van Tendeloo, G.; Liz-Marzán, L. M. Monitoring galvanic replacement through three-dimensional morphological and chemical mapping. *Nano Letters* 2014, *14*, 3220–3226.

129. Shi, L.; Liang, L.; Ma, J.; Wang, F.; Sun, J. Enhanced photocatalytic activity over the Ag_2O–gC_3N_4 composite under visible light. *Catalysis Science & Technology* 2014, *4*, 758–765.
130. Xu, M.; Han, L.; Dong, S. Facile fabrication of highly efficient g-C_3N_4/Ag_2O heterostructured photocatalysts with enhanced visible-light photocatalytic activity. *ACS Applied Materials & Interfaces* 2013, *5*, 12533–12540.
131. Akagi, F.; Matsuo, T.; Kawaguchi, H. Dinitrogen cleavage by a diniobium tetrahydride complex: formation of a nitride and its conversion into imide species. *Angewandte Chemie* 2007, *119*, 8934–8937.
132. Shima, T.; Hu, S.; Luo, G.; Kang, X.; Luo, Y.; Hou, Z. Dinitrogen cleavage and hydrogenation by a trinuclear titanium polyhydride complex. *Science* 2013, *340*, 1549–1552.
133. Wang, J.; Yu, L.; Hu, L.; Chen, G.; Xin, H.; Feng, X. Ambient ammonia synthesis via palladium-catalyzed electrohydrogenation of dinitrogen at low overpotential. *Nature Communications* 2018, *9*, 1795.
134. Kim, K.; Lee, S. J.; Kim, D. Y.; Yoo, C. Y.; Choi, J. W.; Kim, J. N.; Woo, Y.; Yoon, H. C.; Han, J. I. Electrochemical synthesis of ammonia from water and nitrogen: a lithium-mediated approach using lithium-ion conducting glass ceramics. *ChemSusChem* 2018, *11*, 120–124.
135. Wang, P.; Chang, F.; Gao, W.; Guo, J.; Wu, G.; He, T.; Chen, P. Breaking scaling relations to achieve low-temperature ammonia synthesis through LiH-mediated nitrogen transfer and hydrogenation. *Nature Chemistry* 2017, *9*, 64–70.
136. Nørskov, J. K.; Bligaard, T.; Hvolbæk, B.; Abild-Pedersen, F.; Chorkendorff, I.; Christensen, C. H. The nature of the active site in heterogeneous metal catalysis. *Chemical Society Reviews* 2008, *37*, 2163–2171.
137. Nazemi, M.; Padgett, J.; Hatzell, M. C. Acid/base multi-ion exchange membrane-based electrolysis system for water splitting. *Energy Technology* 2017, *5*, 1191–1194.
138. Yi, C.-W.; Luo, K.; Wei, T.; Goodman, D. The composition and structure of Pd– Au surfaces. *The Journal of Physical Chemistry B* 2005, *109*, 18535–18540.
139. Fu, G.-T.; Liu, C.; Zhang, Q.; Chen, Y.; Tang, Y.-W. Polyhedral palladium–silver alloy nanocrystals as highly active and stable electrocatalysts for the formic acid oxidation reaction. *Scientific Reports* 2015, *5*, 13703.
140. Nahar, L.; Farghaly, A. A.; Esteves, R. J. A.; Arachchige, I. U. Shape controlled synthesis of Au/Ag/Pd nanoalloys and their oxidation-induced self-assembly into electrocatalytically active aerogel monoliths. *Chemistry of Materials* 2017, *29*, 7704–7715.
141. Chen, J.; Wiley, B.; McLellan, J.; Xiong, Y.; Li, Z.-Y.; Xia, Y. Optical properties of Pd–Ag and Pt– Ag nanoboxes synthesized via galvanic replacement reactions. *Nano Letters* 2005, *5*, 2058–2062.
142. Nazemi, M.; Soule, L.; Liu, M.; El-Sayed, M. A. Ambient ammonia electrosynthesis from nitrogen and water by incorporating palladium in bimetallic gold–silver nanocages. *Journal of The Electrochemical Society* 2020, *167*, 054511.
143. Verma, S.; Lu, S.; Kenis, P. J. Co-electrolysis of CO_2 and glycerol as a pathway to carbon chemicals with improved technoeconomics due to low electricity consumption. *Nature Energy* 2019, *4*, 466.
144. Soloveichik, G. Renewable energy to fuels through utilization of energy dense liquids (REFUEL). *ARPA-E*. https://arpa-e.energy.gov/?q=arpa-e-programs/refuel. 2016.
145. Andersen, S. Z.; Čolić, V.; Yang, S.; Schwalbe, J. A.; Nielander, A. C.; McEnaney, J. M.; Enemark-Rasmussen, K.; Baker, J. G.; Singh, A. R.; Rohr, B. A. A rigorous electrochemical ammonia synthesis protocol with quantitative isotope measurements. *Nature* 2019, 504–570.

146. Greenlee, L. F.; Renner, J. N.; Foster, S. L. The use of controls for consistent and accurate measurements of electrocatalytic ammonia synthesis from dinitrogen. *ACS Catalysis* 2018, *8*, 7820–7827.
147. Suryanto, B. H.; Du, H.-L.; Wang, D.; Chen, J.; Simonov, A. N.; MacFarlane, D. R. Challenges and prospects in the catalysis of electroreduction of nitrogen to ammonia. *Nature Catalysis* 2019, *2*, 290.
148. Van Tamelen, E. E.; Akermark, B. Electrolytic reduction of molecular nitrogen. *Journal of the American Chemical Society* 1968, *90*, 4492–4493.
149. Yan, Z.; Ji, M.; Xia, J.; Zhu, H. Recent advanced materials for electrochemical and photoelectrochemical synthesis of ammonia from dinitrogen: one step closer to a sustainable energy future. *Advanced Energy Materials* 2020, *10*, 1902020.
150. Marnellos, G.; Stoukides, M. Ammonia synthesis at atmospheric pressure. *Science* 1998, *282*, 98–100.
151. Shipman, M. A.; Symes, M. D. Recent progress towards the electrosynthesis of ammonia from sustainable resources. *Catalysis Today* 2017, *286*, 57–68.
152. Köleli, F.; Röpke, T. Electrochemical hydrogenation of dinitrogen to ammonia on a polyaniline electrode. *Applied Catalysis B: Environmental* 2006, *62*, 306–310.
153. Liu, G.; Cui, Z.; Han, M.; Zhang, S.; Zhao, C.; Chen, C.; Wang, G.; Zhang, H. Ambient electrosynthesis of ammonia on a core–shell-structured Au@ CeO_2 catalyst: contribution of oxygen vacancies in CeO_2. *Chemistry–A European Journal* 2019, *25*, 5904–5911.
154. Suryanto, B. H.; Kang, C. S.; Wang, D.; Xiao, C.; Zhou, F.; Azofra, L. M.; Cavallo, L.; Zhang, X.; MacFarlane, D. R. Rational electrode–electrolyte design for efficient ammonia electrosynthesis under ambient conditions. *ACS Energy Letters* 2018, *3*, 1219–1224.
155. Kim, K.; Yoo, C.-Y.; Kim, J.-N.; Yoon, H. C.; Han, J.-I. Electrochemical synthesis of ammonia from water and nitrogen in ethylenediamine under ambient temperature and pressure. *Journal of the Electrochemical Society* 2016, *163*, F1523.
156. Kim, K.; Lee, N.; Yoo, C.-Y.; Kim, J.-N.; Yoon, H. C.; Han, J.-I. Communication—electrochemical reduction of nitrogen to ammonia in 2-propanol under ambient temperature and pressure. *Journal of the Electrochemical Society* 2016, *163*, F610.
157. Kong, J.; Lim, A.; Yoon, C.; Jang, J. H.; Ham, H. C.; Han, J.; Nam, S.; Kim, D.; Sung, Y.-E.; Choi, J. Electrochemical synthesis of NH_3 at low temperature and atmospheric pressure using a γ-Fe_2O_3 catalyst. *ACS Sustainable Chemistry & Engineering* 2017, *5*, 10986–10995.
158. Yang, X.; Nash, J.; Anibal, J.; Dunwell, M.; Kattel, S.; Stavitski, E.; Attenkofer, K.; Chen, J. G.; Yan, Y.; Xu, B. Mechanistic insights into electrochemical nitrogen reduction reaction on vanadium nitride nanoparticles. *Journal of the American Chemical Society* 2018, *140*, 13387–13391.
159. Ithisuphalap, K.; Zhang, H.; Guo, L.; Yang, Q.; Yang, H.; Wu, G. Photocatalysis and photoelectrocatalysis methods of nitrogen reduction for sustainable ammonia synthesis. *Small Methods* 2019, *3*, 1800352.
160. Ali, M.; Zhou, F.; Chen, K.; Kotzur, C.; Xiao, C.; Bourgeois, L.; Zhang, X.; MacFarlane, D. R. Nanostructured photoelectrochemical solar cell for nitrogen reduction using plasmon-enhanced black silicon. *Nature Communications* 2016, *7*, 1–5.
161. Li, C.; Wang, T.; Zhao, Z. J.; Yang, W.; Li, J. F.; Li, A.; Yang, Z.; Ozin, G. A.; Gong, J. Promoted fixation of molecular nitrogen with surface oxygen vacancies on plasmon-enhanced TiO_2 photoelectrodes. *Angewandte Chemie International Edition* 2018, *57*, 5278–5282.

162. Zhu, D.; Zhang, L.; Ruther, R. E.; Hamers, R. J. Photo-illuminated diamond as a solid-state source of solvated electrons in water for nitrogen reduction. *Nature Materials* 2013, *12*, 836–841.
163. Oshikiri, T.; Ueno, K.; Misawa, H. Plasmon-induced ammonia synthesis through nitrogen photofixation with visible light irradiation. *Angewandte Chemie* 2014, *126*, 9960–9963.
164. Huan, T. N.; Dalla Corte, D. A.; Lamaison, S.; Karapinar, D.; Lutz, L.; Menguy, N.; Foldyna, M.; Turren-Cruz, S.-H.; Hagfeldt, A.; Bella, F. Low-cost high-efficiency system for solar-driven conversion of CO_2 to hydrocarbons. *Proceedings of the National Academy of Sciences* 2019, *116*, 9735–9740.
165. Bullock, J.; Srankó, D. F.; Towle, C. M.; Lum, Y.; Hettick, M.; Scott, M.; Javey, A.; Ager, J. Efficient solar-driven electrochemical CO_2 reduction to hydrocarbons and oxygenates. *Energy & Environmental Science* 2017, *10*, 2222–2230.
166. Corby, S.; Francàs, L.; Kafizas, A.; Durrant, J. R. Determining the role of oxygen vacancies in the photoelectrocatalytic performance of WO_3 for water oxidation. *Chemical Science* 2020, *11*, 2907–2914.
167. Ye, W.; Arif, M.; Fang, X.; Mushtaq, M. A.; Chen, X.; Yan, D. Efficient photoelectrochemical route for the ambient reduction of N_2 to NH_3 based on nanojunctions assembled from MoS_2 nanosheets and TiO_2. *ACS Applied Materials & Interfaces* 2019, *11*, 28809–28817.
168. Schreier, M.; Héroguel, F.; Steier, L.; Ahmad, S.; Luterbacher, J. S.; Mayer, M. T.; Luo, J.; Grätzel, M. Solar conversion of CO_2 to CO using Earth-abundant electrocatalysts prepared by atomic layer modification of CuO. *Nature Energy* 2017, *2*, 17087.
169. Liu, H.; Wei, L.; Liu, F.; Pei, Z.; Shi, J.; Wang, Z.-J.; He, D.; Chen, Y. Homogeneous, heterogeneous, and biological catalysts for electrochemical N_2 reduction toward NH_3 under ambient conditions. *ACS Catalysis* 2019, *9*, 5245–5267.
170. Zheng, J.; Lyu, Y.; Qiao, M.; Wang, R.; Zhou, Y.; Li, H.; Chen, C.; Li, Y.; Zhou, H.; Jiang, S. P. Photoelectrochemical synthesis of ammonia on the aerophilic-hydrophilic heterostructure with 37.8% efficiency. *Chem* 2019, *5*, 617–633.
171. Xue, Z.-H.; Zhang, S.-N.; Lin, Y.-X.; Su, H.; Zhai, G.-Y.; Han, J.-T.; Yu, Q.-Y.; Li, X.-H.; Antonietti, M.; Chen, J.-S. Electrochemical reduction of N_2 into NH_3 by donor–acceptor couples of Ni and Au nanoparticles with a 67.8% Faradaic efficiency. *Journal of the American Chemical Society* 2019, *141*, 14976–14980.
172. Lazouski, N.; Schiffer, Z. J.; Williams, K.; Manthiram, K. Understanding continuous lithium-mediated electrochemical nitrogen reduction. *Joule* 2019, *3*, 1127–1139.
173. Lee, H. K.; Koh, C. S. L.; Lee, Y. H.; Liu, C.; Phang, I. Y.; Han, X.; Tsung, C.-K.; Ling, X. Y. Favoring the unfavored: selective electrochemical nitrogen fixation using a reticular chemistry approach. *Science Advances* 2018, *4*, eaar3208.
174. Akagi, F.; Matsuo, T.; Kawaguchi, H. Dinitrogen cleavage by a diniobium tetrahydride complex: formation of a nitride and its conversion into imide species. *Angewandte Chemie International Edition* 2007, *46*, 8778–8781.
175. Suo, L.; Borodin, O.; Gao, T.; Olguin, M.; Ho, J.; Fan, X.; Luo, C.; Wang, C.; Xu, K. "Water-in-salt" electrolyte enables high-voltage aqueous lithium-ion chemistries. *Science* 2015, *350*, 938–943.
176. Nazemi, M.; Ou, P.; Alabbady, A.; Soule, L.; Liu, A.; Song, J.; Sulchek, T. A.; Liu, M.; El-Sayed, M. A. Electrosynthesis of ammonia using porous bimetallic Pd–Ag nanocatalysts in liquid-and gas-phase systems. *ACS Catalysis* 2020, *10*, 10197–10206.
177. Mahmoud, M. A.; O'Neil, D.; El-Sayed, M. A. Hollow and solid metallic nanoparticles in sensing and in nanocatalysis. *Chemistry of Materials* 2014, *26*, 44–58.

178. Valencia, F. J.; Gonzláez, R. I.; Vega, H.; Ruestes, C.; Rogan, J.; Valdivia, J. A.; Bringa, E. M.; Kiwi, M. Mechanical properties obtained by indentation of hollow Pd nanoparticles. *The Journal of Physical Chemistry C* 2018, *122*, 25035–25042.
179. Idrissi, H.; Wang, B.; Colla, M. S.; Raskin, J. P.; Schryvers, D.; Pardoen, T. Ultrahigh strain hardening in thin palladium films with nanoscale twins. *Advanced Materials* 2011, *23*, 2119–2122.
180. Higgins, D.; Hahn, C.; Xiang, C.; Jaramillo, T. F.; Weber, A. Z. Gas-diffusion electrodes for carbon dioxide reduction: a new paradigm. *ACS Energy Letters* 2018, *4*, 317–324.
181. Weng, L.-C.; Bell, A. T.; Weber, A. Z. Towards membrane-electrode assembly systems for CO_2 reduction: a modeling study. *Energy & Environmental Science* 2019, *12*, 1950–1968.
182. Larrazábal, G. O.; Strøm-Hansen, P.; Heli, J. P.; Zeiter, K.; Therkildsen, K. T.; Chorkendorff, I.; Seger, B. Analysis of mass flows and membrane cross-over in CO_2 reduction at high current densities in an MEA-type electrolyzer. *ACS Applied Materials & Interfaces* 2019, *11*, 41281–41288.
183. Schmidt, T.; Gasteiger, H.; Stäb, G.; Urban, P.; Kolb, D.; Behm, R. Characterization of high-surface-area electrocatalysts using a rotating disk electrode configuration. *Journal of the Electrochemical Society* 1998, *145*, 2354–2358.
184. Nazemi, M.; El-Sayed, M. A. Plasmon-enhanced photo (electro) chemical nitrogen fixation under ambient conditions using visible light responsive hybrid hollow Au-Ag_2O nanocages. *Nano Energy* 2019, *63*, 103886.
185. Wang, J.; Yu, L.; Hu, L.; Chen, G.; Xin, H.; Feng, X. Ambient ammonia synthesis via palladium-catalyzed electrohydrogenation of dinitrogen at low overpotential. *Nature Communications* 2018, *9*, 1–7.
186. Tong, W.; Huang, B.; Wang, P.; Li, L.; Shao, Q.; Huang, X. Crystal-Phase-engineered PdCu electrocatalyst for enhanced ammonia synthesis. *Angewandte Chemie International Edition* 2020, *59*, 2649–2653.
187. Wang, H.; Li, Y.; Li, C.; Deng, K.; Wang, Z.; Xu, Y.; Li, X.; Xue, H.; Wang, L. One-pot synthesis of bi-metallic PdRu tripods as an efficient catalyst for electrocatalytic nitrogen reduction to ammonia. *Journal of Materials Chemistry A* 2019, *7*, 801–805.
188. Wang, Z.; Li, C.; Deng, K.; Xu, Y.; Xue, H.; Li, X.; Wang, L.; Wang, H. Ambient nitrogen reduction to ammonia electrocatalyzed by bimetallic PdRu porous nanostructures. *ACS Sustainable Chemistry & Engineering* 2018, *7*, 2400–2405.
189. Wang, H.; Yang, D.; Liu, S.; Yin, S.; Xu, Y.; Li, X.; Wang, Z.; Wang, L. Metal–nonmetal one-dimensional electrocatalyst: AuPdP nanowires for ambient nitrogen reduction to ammonia. *ACS Sustainable Chemistry & Engineering* 2019, *7*, 15772–15777.
190. Kumar, R. D.; Wang, Z.; Li, C.; Kumar, A. V. N.; Xue, H.; Xu, Y.; Li, X.; Wang, L.; Wang, H. Trimetallic PdCuIr with long-spined sea-urchin-like morphology for ambient electroreduction of nitrogen to ammonia. *Journal of Materials Chemistry A* 2019, *7*, 3190–3196.
191. Pang, F.; Wang, Z.; Zhang, K.; He, J.; Zhang, W.; Guo, C.; Ding, Y. Bimodal nanoporous Pd3Cu1 alloy with restrained hydrogen evolution for stable and high yield electrochemical nitrogen reduction. *Nano Energy* 2019, *58*, 834–841.
192. Bogaerts, A.; Neyts, E. C. Plasma technology: an emerging technology for energy storage. *ACS Energy Letters* 2018, *3*, 1013–1027.
193. Fridman, A. *Plasma chemistry*. Cambridge University Press; 2008.
194. Rouwenhorst, K. H.; Kim, H.-H.; Lefferts, L. Vibrationally excited activation of N_2 in plasma-enhanced catalytic ammonia synthesis: a kinetic analysis. *ACS Sustainable Chemistry & Engineering* 2019, *7*, 17515–17522.

195. Carreon, M. L. Plasma catalytic ammonia synthesis: state of the art and future directions. *Journal of Physics D: Applied Physics* 2019, *52*, 483001.
196. Peng, P.; Chen, P.; Schiappacasse, C.; Zhou, N.; Anderson, E.; Chen, D.; Liu, J.; Cheng, Y.; Hatzenbeller, R.; Addy, M. A review on the non-thermal plasma-assisted ammonia synthesis technologies. *Journal of Cleaner Production* 2018, *177*, 597–609.
197. Kubota, Y.; Koga, K.; Ohno, M.; Hara, T. Synthesis of ammonia through direct chemical reactions between an atmospheric nitrogen plasma jet and a liquid. *Plasma and Fusion Research* 2010, *5*, 42.
198. Haruyama, T.; Namise, T.; Shimoshimizu, N.; Uemura, S.; Takatsuji, Y.; Hino, M.; Yamasaki, R.; Kamachi, T.; Kohno, M. Non-catalyzed one-step synthesis of ammonia from atmospheric air and water. *Green Chemistry* 2016, *18*, 4536–4541.
199. Sakakura, T.; Uemura, S.; Hino, M.; Kiyomatsu, S.; Takatsuji, Y.; Yamasaki, R.; Morimoto, M.; Haruyama, T. Excitation of H_2O at the plasma/water interface by UV irradiation for the elevation of ammonia production. *Green Chemistry* 2018, *20*, 627–633.
200. Peng, P.; Chen, P.; Addy, M.; Cheng, Y.; Zhang, Y.; Anderson, E.; Zhou, N.; Schiappacasse, C.; Hatzenbeller, R.; Fan, L. In situ plasma-assisted atmospheric nitrogen fixation using water and spray-type jet plasma. *Chemical Communications* 2018, *54*, 2886–2889.
201. Akay, G.; Zhang, K. Process intensification in ammonia synthesis using novel coassembled supported microporous catalysts promoted by nonthermal plasma. *Industrial & Engineering Chemistry Research* 2017, *56*, 457–468.
202. Snoeckx, R.; Bogaerts, A. Plasma technology–a novel solution for CO_2 conversion? *Chemical Society Reviews* 2017, *46*, 5805–5863.
203. Kim, H.-H.; Teramoto, Y.; Ogata, A.; Takagi, H.; Nanba, T. Plasma catalysis for environmental treatment and energy applications. *Plasma Chemistry and Plasma Processing* 2016, *36*, 45–72.
204. Mehta, P.; Barboun, P.; Herrera, F. A.; Kim, J.; Rumbach, P.; Go, D. B.; Hicks, J. C.; Schneider, W. F. Overcoming ammonia synthesis scaling relations with plasma-enabled catalysis. *Nature Catalysis* 2018, *1*, 269–275.
205. Patel, H.; Sharma, R. K.; Kyriakou, V.; Pandiyan, A.; Welzel, S.; van de Sanden, M. C.; Tsampas, M. N. Plasma-activated electrolysis for cogeneration of nitric oxide and hydrogen from water and nitrogen. *ACS Energy Letters* 2019, *4*, 2091–2095.
206. Kumari, S.; Pishgar, S.; Schwarting, M. E.; Paxton, W. F.; Spurgeon, J. M. Synergistic plasma-assisted electrochemical reduction of nitrogen to ammonia. *Chemical Communications* 2018, *54*, 13347–13350.
207. Arora, P.; Sharma, I.; Hoadley, A.; Mahajani, S.; Ganesh, A. Remote, small-scale, 'greener' routes of ammonia production. *Journal of Cleaner Production* 2018, *199*, 177–192.
208. Dhar, N.; Pant, N. Nitrogen loss from soils and oxide surfaces. *Nature* 1944, *153*, 115–116.
209. Schrauzer, G.; Guth, T. Photolysis of water and photoreduction of nitrogen on titanium dioxide. *Journal of the American Chemical Society* 2002, *99*, 7189–7193.
210. Li, Q.-S.; Domen, K.; Naito, S.; Onishi, T.; Tamaru, K. Photocatalytic synthesis and photodecomposition of ammonia over $SrTiO_3$ and $BaTiO_3$ based catalysts. *Chemistry Letters* 1983, *12*, 321–324.
211. Sun, S.; Li, X.; Wang, W.; Zhang, L.; Sun, X. Photocatalytic robust solar energy reduction of dinitrogen to ammonia on ultrathin MoS_2. *Applied Catalysis B: Environmental* 2017, *200*, 323–329.

212. Zhang, S.; Zhao, Y.; Shi, R.; Zhou, C.; Waterhouse, G. I.; Wu, L. Z.; Tung, C. H.; Zhang, T. Efficient photocatalytic nitrogen fixation over Cuδ+-modified defective ZnAl-layered double hydroxide nanosheets. *Advanced Energy Materials* 2020, *10*, 1901973.
213. Kudo, A.; Miseki, Y. Heterogeneous photocatalyst materials for water splitting. *Chemical Society Reviews* 2009, *38*, 253–278.
214. Li, H.; Shang, J.; Shi, J.; Zhao, K.; Zhang, L. Facet-dependent solar ammonia synthesis of BiOCl nanosheets via a proton-assisted electron transfer pathway. *Nanoscale* 2016, *8*, 1986–1993.
215. Shi, R.; Zhao, Y.; Waterhouse, G. I.; Zhang, S.; Zhang, T. Defect engineering in photocatalytic nitrogen fixation. *ACS Catalysis* 2019, *9*, 9739–9750.
216. Cao, N.; Chen, Z.; Zang, K.; Xu, J.; Zhong, J.; Luo, J.; Xu, X.; Zheng, G. Doping strain induced bi-Ti_3+ pairs for efficient N_2 activation and electrocatalytic fixation. *Nature Communications* 2019, *10*, 1–12.
217. Zhao, Y.; Zhao, Y.; Shi, R.; Wang, B.; Waterhouse, G. I.; Wu, L. Z.; Tung, C. H.; Zhang, T. Tuning oxygen vacancies in ultrathin TiO_2 nanosheets to boost photocatalytic nitrogen fixation up to 700 nm. *Advanced Materials* 2019, *31*, 1806482.
218. Zhao, Y.; Zhao, Y.; Waterhouse, G. I.; Zheng, L.; Cao, X.; Teng, F.; Wu, L. Z.; Tung, C. H.; O'Hare, D.; Zhang, T. Layered-double-hydroxide nanosheets as efficient visible-light-driven photocatalysts for dinitrogen fixation. *Advanced Materials* 2017, *29*, 1703828.
219. Hirakawa, H.; Hashimoto, M.; Shiraishi, Y.; Hirai, T. Photocatalytic conversion of nitrogen to ammonia with water on surface oxygen vacancies of titanium dioxide. *Journal of the American Chemical Society* 2017, *139*, 10929–10936.
220. Yuan, S.-J.; Chen, J.-J.; Lin, Z.-Q.; Li, W.-W.; Sheng, G.-P.; Yu, H.-Q. Nitrate formation from atmospheric nitrogen and oxygen photocatalysed by nano-sized titanium dioxide. *Nature Communications* 2013, *4*, 1–7.
221. Liu, Y.; Cheng, M.; He, Z.; Gu, B.; Xiao, C.; Zhou, T.; Guo, Z.; Liu, J.; He, H.; Ye, B. Pothole-rich ultrathin WO_3 nanosheets that trigger N≡ N bond activation of nitrogen for direct nitrate photosynthesis. *Angewandte Chemie International Edition* 2019, *58*, 731–735.
222. Linic, S.; Christopher, P.; Ingram, D. B. Plasmonic-metal nanostructures for efficient conversion of solar to chemical energy. *Nature Materials* 2011, *10*, 911–921.
223. Schrauzer, G. N. Photoreduction of nitrogen on TiO_2 and TiO_2-containing minerals. In *Energy efficiency and renewable energy through nanotechnology*. Springer; 2011, pp 601–623.
224. Rusina, O.; Linnik, O.; Eremenko, A.; Kisch, H. Nitrogen photofixation on nanostructured iron titanate films. *Chemistry–A European Journal* 2003, *9*, 561–565.
225. Rusina, O.; Eremenko, A.; Frank, G.; Strunk, H. P.; Kisch, H. Nitrogen photofixation at nanostructured iron titanate films. *Angewandte Chemie International Edition* 2001, *40*, 3993–3995.
226. Chen, X.; Li, N.; Kong, Z.; Ong, W.-J.; Zhao, X. Photocatalytic fixation of nitrogen to ammonia: state-of-the-art advancements and future prospects. *Materials Horizons* 2018, *5*, 9–27.
227. Li, H.; Shang, J.; Ai, Z.; Zhang, L. Efficient visible light nitrogen fixation with BiOBr nanosheets of oxygen vacancies on the exposed {001} facets. *Journal of the American Chemical Society* 2015, *137*, 6393–6399.
228. Li, J.; Li, H.; Zhan, G.; Zhang, L. Solar water splitting and nitrogen fixation with layered bismuth oxyhalides. *Accounts of Chemical Research* 2017, *50*, 112–121.

229. Xue, X.; Chen, R.; Chen, H.; Hu, Y.; Ding, Q.; Liu, Z.; Ma, L.; Zhu, G.; Zhang, W.; Yu, Q. Oxygen vacancy engineering promoted photocatalytic ammonia synthesis on ultrathin two-dimensional bismuth oxybromide nanosheets. *Nano Letters* 2018, *18*, 7372–7377.
230. Boriskina, S. V.; Ghasemi, H.; Chen, G. Plasmonic materials for energy: from physics to applications. *Materials Today* 2013, *16*, 375–386.
231. Saha, K.; Agasti, S. S.; Kim, C.; Li, X.; Rotello, V. M. Gold nanoparticles in chemical and biological sensing. *Chemical Reviews* 2012, *112*, 2739–2779.
232. Murray, W. A.; Barnes, W. L. Plasmonic materials. *Advanced Materials* 2007, *19*, 3771–3782.
233. Clavero, C. Plasmon-induced hot-electron generation at nanoparticle/metal-oxide interfaces for photovoltaic and photocatalytic devices. *Nature Photonics* 2014, *8*, 95–103.
234. Hu, C.; Chen, X.; Jin, J.; Han, Y.; Chen, S.; Ju, H.; Cai, J.; Qiu, Y.; Gao, C.; Wang, C. Surface plasmon enabling nitrogen fixation in pure water through a dissociative mechanism under mild conditions. *Journal of the American Chemical Society* 2019, *141*, 7807–7814.
235. Gupta, M. K.; König, T.; Near, R.; Nepal, D.; Drummy, L. F.; Biswas, S.; Naik, S.; Vaia, R. A.; El-Sayed, M. A.; Tsukruk, V. V. Surface assembly and plasmonic properties in strongly coupled segmented gold nanorods. *Small* 2013, *9*, 2979–2990.
236. Yang, J.; Guo, Y.; Jiang, R.; Qin, F.; Zhang, H.; Lu, W.; Wang, J.; Yu, J. C. High-efficiency "working-in-tandem" nitrogen photofixation achieved by assembling plasmonic gold nanocrystals on ultrathin titania nanosheets. *Journal of the American Chemical Society* 2018, *140*, 8497–8508.
237. Shi, E.; Gao, Y.; Finkenauer, B. P.; Coffey, A. H.; Dou, L. Two-dimensional halide perovskite nanomaterials and heterostructures. *Chemical Society Reviews* 2018, *47*, 6046–6072.
238. Cheng, H.; Huang, B.; Dai, Y. Engineering BiOX (X= Cl, Br, I) nanostructures for highly efficient photocatalytic applications. *Nanoscale* 2014, *6*, 2009–2026.
239. Akolekar, D. B.; Bhargava, S. K.; Foran, G. EXAFS studies on gold nanoparticles over novel catalytic materials. *Radiation Physics and Chemistry* 2006, *75*, 1948–1952.
240. Kafizas, A.; Parry, S. A.; Chadwick, A. V.; Carmalt, C. J.; Parkin, I. P. An EXAFS study on the photo-assisted growth of silver nanoparticles on titanium dioxide thin-films and the identification of their photochromic states. *Physical Chemistry Chemical Physics* 2013, *15*, 8254–8263.
241. Chao, Y.; Zhou, P.; Li, N.; Lai, J.; Yang, Y.; Zhang, Y.; Tang, Y.; Yang, W.; Du, Y.; Su, D. Ultrathin visible-light-driven Mo incorporating $In2O_3$–$ZnIn2Se_4$ Z-scheme nanosheet photocatalysts. *Advanced Materials* 2019, *31*, 1807226.
242. Xu, C.; Qiu, P.; Li, L.; Chen, H.; Jiang, F.; Wang, X. Bismuth subcarbonate with designer defects for broad-spectrum photocatalytic nitrogen fixation. *ACS Applied Materials & Interfaces* 2018, *10*, 25321–25328.
243. Kumar, D. P.; Kim, E. H.; Park, H.; Chun, S. Y.; Gopannagari, M.; Bhavani, P.; Reddy, D. A.; Song, J. K.; Kim, T. K. Tuning band alignments and charge-transport properties through $MoSe_2$ bridging between MoS_2 and cadmium sulfide for enhanced hydrogen production. *ACS Applied Materials & Interfaces* 2018, *10*, 26153–26161.
244. Zhang, N.; Jalil, A.; Wu, D.; Chen, S.; Liu, Y.; Gao, C.; Ye, W.; Qi, Z.; Ju, H.; Wang, C. Refining defect states in W18O49 by Mo doping: a strategy for tuning N_2 activation towards solar-driven nitrogen fixation. *Journal of the American Chemical Society* 2018, *140*, 9434–9443.
245. Shiraishi, Y.; Shiota, S.; Kofuji, Y.; Hashimoto, M.; Chishiro, K.; Hirakawa, H.; Tanaka, S.; Ichikawa, S.; Hirai, T. Nitrogen fixation with water on carbon-nitride-based metal-free photocatalysts with 0.1% solar-to-ammonia energy conversion efficiency. *ACS Applied Energy Materials* 2018, *1*, 4169–4177.

246. Emerson, K.; Russo, R. C.; Lund, R. E.; Thurston, R. V. Aqueous ammonia equilibrium calculations: effect of pH and temperature. *Journal of the Fisheries Board of Canada* 1975, *32*, 2379–2383.
247. Körner, S.; Das, S. K.; Veenstra, S.; Vermaat, J. E. The effect of pH variation at the ammonium/ammonia equilibrium in wastewater and its toxicity to Lemna gibba. *Aquatic Botany* 2001, *71*, 71–78.
248. Shah, S. B.; Westerman, P. W.; Arogo, J. Measuring ammonia concentrations and emissions from agricultural land and liquid surfaces: a review. *Journal of the Air & Waste Management Association* 2006, *56*, 945–960.
249. Yan, D.; Li, H.; Chen, C.; Zou, Y.; Wang, S. Defect engineering strategies for nitrogen reduction reactions under ambient conditions. *Small Methods* 2019, *3*, 1800331.
250. Watt, G. W.; Chrisp, J. D. Spectrophotometric method for determination of hydrazine. *Analytical Chemistry* 1952, *24*, 2006–2008.
251. Vanselow, A. Preparation of Nessler's reagent. *Industrial & Engineering Chemistry Analytical Edition* 1940, *12*, 516–517.
252. Zhou, J.-H.; Zhang, Y.-W. Metal-based heterogeneous electrocatalysts for reduction of carbon dioxide and nitrogen: mechanisms, recent advances and perspective. *Reaction Chemistry & Engineering* 2018, *3*, 591–625.
253. Krug, F.; Růžička, J.; Hansen, E. Determination of ammonia in low concentrations with Nessler's reagent by flow injection analysis. *Analyst* 1979, *104*, 47–54.
254. Thompson, J.; Morrison, G. Determination of organic nitrogen. Control of variables in the use of Nessler's reagent. *Analytical Chemistry* 1951, *23*, 1153–1157.
255. Yuen, S.; Pollard, A. The determination of nitrogen in agricultural materials by the nessler reagent. I.—preparation of the reagent. *Journal of the Science of Food and Agriculture* 1952, *3*, 441–447.
256. Crosby, N. Determination of ammonia by the Nessler method in waters containing hydrazine. *Analyst* 1968, *93*, 406–408.
257. Gao, X.; Wen, Y.; Qu, D.; An, L.; Luan, S.; Jiang, W.; Zong, X.; Liu, X.; Sun, Z. Interference effect of alcohol on Nessler's reagent in photocatalytic nitrogen fixation. *ACS Sustainable Chemistry & Engineering* 2018, *6*, 5342–5348.
258. Zhou, L.; Boyd, C. E. Comparison of Nessler, phenate, salicylate and ion selective electrode procedures for determination of total ammonia nitrogen in aquaculture. *Aquaculture* 2016, *450*, 187–193.
259. Rice, E.; Baird, R.; Eaton, A.; Clesceri, L. Standard methods fort he examination of water and wastewater. *Part* 2012, *2540*, 2–66.
260. Hansen, H.; Koroleff, F. *Determination of nutrients: methods of seawater analysis*. Eds. K. Grasshoff; K. Kremling; M. Ernhardt; Wiley-VCH; 1999.
261. Searle, P. L. The Berthelot or indophenol reaction and its use in the analytical chemistry of nitrogen: a review. *Analyst* 1984, *109*, 549–568.
262. Bolleter, W.; Bushman, C.; Tidwell, P. W. Spectrophotometric determination of ammonia as indophenol. *Analytical Chemistry* 1961, *33*, 592–594.
263. Harfmann, R.; Crouch, S. Kinetic study of Berthelot reaction steps in the absence and presence of coupling reagents. *Talanta* 1989, *36*, 261–269.
264. Hampson, B. The analysis of ammonia in polluted sea water. *Water Research* 1977, *11*, 305–308.
265. Ivančič, I.; Degobbis, D. An optimal manual procedure for ammonia analysis in natural waters by the indophenol blue method. *Water Research* 1984, *18*, 1143–1147.
266. Solorzano, L. Determination of ammonia in natural waters by the phenolhypochlorite method 1.1. This research was fully supported by US Atomic Energy Commission Contract No. ATS (11-1) GEN 10, PA 20. *Limnology and Oceanography* 1969, *14*, 799–801.

267. Grasshoff, K.; Johannsen, H. A new sensitive and direct method for the automatic determination of ammonia in sea water. *ICES Journal of Marine Science* 1972, *34*, 516–521.
268. Tzollas, N. M.; Zachariadis, G. A.; Anthemidis, A. N.; Stratis, J. A. A new approach to indophenol blue method for determination of ammonium in geothermal waters with high mineral content. *International Journal of Environmental and Analytical Chemistry* 2010, *90*, 115–126.
269. Small, H. *Ion chromatography*. Springer Science & Business Media; 2013.
270. Michalski, R. Ion chromatography applications in wastewater analysis. *Separations* 2018, *5*, 16.
271. Michalski, R.; Kurzyca, I. Determination of nitrogen species (nitrate, nitrite and ammonia ions) in environmental samples by ion chromatography. *Polish Journal of Environmental Studies* 2006, *15*, 5–18.
272. Michalski, R. Applications of ion chromatography for the determination of inorganic cations. *Critical Reviews in Analytical Chemistry* 2009, *39*, 230–250.
273. Thomas, D.; Rey, M.; Jackson, P. Determination of inorganic cations and ammonium in environmental waters by ion chromatography with a high-capacity cation-exchange column. *Journal of Chromatography A* 2002, *956*, 181–186.
274. Bruzzoniti, M. C.; De Carlo, R. M.; Fungi, M. Simultaneous determination of alkali, alkaline earths and ammonium in natural waters by ion chromatography. *Journal of Separation Science* 2008, *31*, 3182–3189.
275. Rey, M. A.; Pohl, C. A.; Jagodzinski, J. J.; Kaiser, E. Q.; Riviello, J. M. A new approach to dealing with high-to-low concentration ratios of sodium and ammonium ions in ion chromatography. *Journal of Chromatography A* 1998, *804*, 201–209.
276. Rey, M. A.; Riviello, J. M.; Pohl, C. A. Column switching for difficult cation separations. *Journal of Chromatography A* 1997, *789*, 149–155.
277. Dabundo, R.; Lehmann, M. F.; Treibergs, L.; Tobias, C. R.; Altabet, M. A.; Moisander, P. H.; Granger, J. The contamination of commercial $15N_2$ gas stocks with 15N–labeled nitrate and ammonium and consequences for nitrogen fixation measurements. *PloS One* 2014, *9*, e110335.
278. Chen, G. F.; Ren, S.; Zhang, L.; Cheng, H.; Luo, Y.; Zhu, K.; Ding, L. X.; Wang, H. Advances in electrocatalytic N_2 reduction—strategies to tackle the selectivity challenge. *Small Methods* 2019, *3*, 1800337.
279. Choi, J.; Suryanto, B. H.; Wang, D.; Du, H.-L.; Hodgetts, R. Y.; Vallana, F. M. F.; MacFarlane, D. R.; Simonov, A. N. Identification and elimination of false positives in electrochemical nitrogen reduction studies. *Nature Communications* 2020, *11*, 1–10.
280. Skulason, E.; Bligaard, T.; Gudmundsdóttir, S.; Studt, F.; Rossmeisl, J.; Abild-Pedersen, F.; Vegge, T.; Jonsson, H.; Nørskov, J. K. A theoretical evaluation of possible transition metal electro-catalysts for N_2 reduction. *Physical Chemistry Chemical Physics* 2012, *14*, 1235–1245.
281. Wang, S.; Petzold, V.; Tripkovic, V.; Kleis, J.; Howalt, J. G.; Skulason, E.; Fernandez, E.; Hvolbæk, B.; Jones, G.; Toftelund, A. Universal transition state scaling relations for (de) hydrogenation over transition metals. *Physical Chemistry Chemical Physics* 2011, *13*, 20760–20765.
282. Seh, Z. W.; Kibsgaard, J.; Dickens, C. F.; Chorkendorff, I.; Nørskov, J. K.; Jaramillo, T. F. Combining theory and experiment in electrocatalysis: insights into materials design. *Science* 2017, *355*, eaad4998.
283. Medford, A. J.; Vojvodic, A.; Hummelshøj, J. S.; Voss, J.; Abild-Pedersen, F.; Studt, F.; Bligaard, T.; Nilsson, A.; Nørskov, J. K. From the Sabatier principle to a predictive theory of transition-metal heterogeneous catalysis. *Journal of Catalysis* 2015, *328*, 36–42.

284. Rod, T. H.; Logadottir, A.; Nørskov, J. K. Ammonia synthesis at low temperatures. *The Journal of Chemical Physics* 2000, *112*, 5343–5347.
285. Back, S.; Jung, Y. On the mechanism of electrochemical ammonia synthesis on the Ru catalyst. *Physical Chemistry Chemical Physics* 2016, *18*, 9161–9166.
286. Rosen, B. A.; Salehi-Khojin, A.; Thorson, M. R.; Zhu, W.; Whipple, D. T.; Kenis, P. J.; Masel, R. I. Ionic liquid–mediated selective conversion of CO_2 to CO at low overpotentials. *Science* 2011, *334*, 643–644.
287. Zhang, L.; Sharada, S. M.; Singh, A. R.; Rohr, B. A.; Su, Y.; Qiao, L.; Nørskov, J. K. A theoretical study of the effect of a non-aqueous proton donor on electrochemical ammonia synthesis. *Physical Chemistry Chemical Physics* 2018, *20*, 4982–4989.
288. Malkani, A. S.; Anibal, J.; Chang, X.; Xu, B. Bridging the gap in the mechanistic understanding of electrocatalysis via in-situ characterizations. *Iscience* 2020, *23*, 101776.
289. Kim, Y.-G.; Baricuatro, J. H.; Javier, A.; Gregoire, J. M.; Soriaga, M. P. The evolution of the polycrystalline copper surface, first to Cu (111) and then to Cu (100), at a fixed CO_2RR potential: a study by operando EC-STM. *Langmuir* 2014, *30*, 15053–15056.
290. Zhao, Y.; Chang, X.; Malkani, A. S.; Yang, X.; Thompson, L.; Jiao, F.; Xu, B. Speciation of Cu surfaces during the electrochemical CO reduction reaction. *Journal of the American Chemical Society* 2020, *142*, 9735–9743.
291. Zandi, O.; Hamann, T. W. Determination of photoelectrochemical water oxidation intermediates on haematite electrode surfaces using operando infrared spectroscopy. *Nature Chemistry* 2016, *8*, 778.
292. Deng, Y.; Yeo, B. S. Characterization of electrocatalytic water splitting and CO_2 reduction reactions using in situ/operando Raman spectroscopy. *ACS Catalysis* 2017, *7*, 7873–7889.
293. Deng, Y.; Handoko, A. D.; Du, Y.; Xi, S.; Yeo, B. S. In situ Raman spectroscopy of copper and copper oxide surfaces during electrochemical oxygen evolution reaction: identification of CuIII oxides as catalytically active species. *ACS Catalysis* 2016, *6*, 2473–2481.
294. Yao, Y.; Zhu, S.; Wang, H.; Li, H.; Shao, M. A spectroscopic study on the nitrogen electrochemical reduction reaction on gold and platinum surfaces. *Journal of the American Chemical Society* 2018, *140*, 1496–1501.
295. Fortes, A. D.; Wood, I. G.; Alfè, D.; Hernández, E. R.; Gutmann, M. J.; Sparkes, H. A. Structure, hydrogen bonding and thermal expansion of ammonium carbonate monohydrate. *Acta Crystallographica Section B: Structural Science, Crystal Engineering and Materials* 2014, *70*, 948–962.
296. Carvalho, O. Q.; Adiga, P.; Murthy, S. K.; Fulton, J. L.; Gutiérrez, O. Y.; Stoerzinger, K. A. Understanding the role of surface heterogeneities in electrosynthesis reactions. *Iscience* 2020, *23*, 101814.
297. Zhang, C.; Grass, M. E.; Yu, Y.; Gaskell, K. J.; DeCaluwe, S. C.; Chang, R.; Jackson, G. S.; Hussain, Z.; Bluhm, H.; Eichhorn, B. W. Multielement activity mapping and potential mapping in solid oxide electrochemical cells through the use of operando XPS. *Acs Catalysis* 2012, *2*, 2297–2304.
298. Yu, Y.; Mao, B.; Geller, A.; Chang, R.; Gaskell, K.; Liu, Z.; Eichhorn, B. W. CO_2 activation and carbonate intermediates: an operando AP-XPS study of CO_2 electrolysis reactions on solid oxide electrochemical cells. *Physical Chemistry Chemical Physics* 2014, *16*, 11633–11639.
299. Nellist, M. R.; Laskowski, F. A.; Qiu, J.; Hajibabaei, H.; Sivula, K.; Hamann, T. W.; Boettcher, S. W. Potential-sensing electrochemical atomic force microscopy for in operando analysis of water-splitting catalysts and interfaces. *Nature Energy* 2018, *3*, 46–52.

300. Pfisterer, J. H.; Liang, Y.; Schneider, O.; Bandarenka, A. S. Direct instrumental identification of catalytically active surface sites. *Nature* 2017, *549*, 74–77.
301. Liang, Y.; Csoklich, C.; McLaughlin, D.; Schneider, O.; Bandarenka, A. S. Revealing active sites for hydrogen evolution at Pt and Pd atomic layers on Au surfaces. *ACS Applied Materials & Interfaces* 2019, *11*, 12476–12480.
302. Mariano, R. G.; McKelvey, K.; White, H. S.; Kanan, M. W. Selective increase in CO_2 electroreduction activity at grain-boundary surface terminations. *Science* 2017, *358*, 1187–1192.
303. Meier, J.; Friedrich, K.; Stimming, U. Novel method for the investigation of single nanoparticle reactivity. *Faraday Discussions* 2002, *121*, 365–372.
304. Sun, T.; Yu, Y.; Zacher, B. J.; Mirkin, M. V. Scanning electrochemical microscopy of individual catalytic nanoparticles. *Angewandte Chemie International Edition* 2014, *53*, 14120–14123.
305. Wang, M.; Árnadóttir, L.; Xu, Z. J.; Feng, Z. In situ X-ray absorption spectroscopy studies of nanoscale electrocatalysts. *Nano-Micro Letters* 2019, *11*, 47.
306. Mukerjee, S.; McBreen, J. Effect of particle size on the electrocatalysis by carbon-supported Pt electrocatalysts: an in situ XAS investigation. *Journal of Electroanalytical Chemistry* 1998, *448*, 163–171.
307. Outlook, E. A. E. Technical report. *US Energy Information Administration*. https://www.eia.gov/outlooks/ieo/pdf/ieo2019.pdf. 2019.

Index

O

P

Q

R

S

T

V

W